Scania Trucks: Pioneering the Future of Transport

Etienne Psaila

Scania Trucks: Pioneering the Future of Transport

Copyright © 2025 by Etienne Psaila. All rights reserved.

First Edition: **January 2025**

No part of this publication may be reproduced, distributed, or transmitted in any form or by any means, including photocopying, recording, or other electronic or mechanical methods, without the prior written permission of the publisher, except in the case of brief quotations embodied in critical reviews and certain other non-commercial uses permitted by copyright law.

ISBN: 978-9918-629-00-8

Table of Contents

Chapter 1: The Birth of Scania

Chapter 2: Scania-Vabis Takes the Lead

Chapter 3: Scania in the Pre-War Years

Chapter 4: World War II and Post-War Rebuilding

Chapter 5: The 1950s and 1960s: A New Age of Heavy-Duty Trucks

Chapter 6: Scania in the 1970s: Innovation and Global Expansion

Chapter 7: The 1980s: A Decade of Transformation

Chapter 8: The 1990s: Emissions, Efficiency, and the Modular System

Chapter 9: Entering the 21st Century: Scania's Focus on Sustainability

Chapter 10: Scania's Digital Revolution

Chapter 11: Scania's Electric Future: Green Transport and Innovation

Chapter 12: The Role of Scania in the Global Supply Chain

Chapter 13: The Driver's Experience: Safety, Comfort, and Technology

Chapter 14: Scania's Legacy in Motorsports and Performance

Chapter 15: Scania Today: A Global Leader in Heavy Transport

Chapter 16: The Road Ahead: Scania's Vision for 2030 and Beyond

Chapter 1: The Birth of Scania

The early 20th century in Sweden was a time of rapid industrialization, where new technologies were beginning to reshape the way people lived and worked. The country's landscape, with its vast forests, expansive plains, and harsh winters, posed unique challenges for transportation. In response, Swedish engineers and entrepreneurs began to explore new methods of moving goods and people across the country's rugged terrain. This was the fertile ground in which two pioneering companies, Vabis and Scania, would take their first steps toward shaping the future of transportation.

Early Transportation Developments in Sweden

Sweden, like much of Europe, was still largely dependent on horse-drawn carriages and carts for moving freight at the turn of the century. But as the automobile industry began to take hold elsewhere in the world, Sweden's inventors and engineers started to look for ways to introduce mechanized transport to the Swedish landscape. The country's roads were not well-suited for the heavy vehicles that were starting to be developed, and the weather, especially during long, harsh winters, presented an additional set of challenges.

Yet, Sweden was no stranger to ingenuity. Swedish engineers, recognizing the potential for motorized transport, began experimenting with early automobiles and trucks. The introduction of internal combustion engines gave rise to an entirely new way of moving freight across the country's varied geography. However, the limited infrastructure and high demand for reliable vehicles made it difficult for businesses to transition away from traditional methods. That is when two companies—Vabis and Scania—emerged, each with its own vision of what transportation in Sweden should become.

The Founding of Vabis and Scania

In the heart of Sweden, in Södertälje, the Vabis company was born. Vabis (Vagnfabriks Aktiebolaget i Södertälje), established in 1891, initially produced railway wagons before shifting its focus to automobiles. Its founders, a group of Swedish entrepreneurs, saw the potential in moving away from traditional rail transport and embraced the growing automotive industry. They quickly began designing and manufacturing vehicles that were tailored to Swedish conditions—strong enough to handle the country's varied terrains, yet efficient and durable enough for industrial needs.

At the same time, in the southern part of Sweden, the Scania company was taking its first steps. Founded in 1900 in Malmö by a group of Swedish businessmen, Scania initially focused on manufacturing bicycles and, later, motorized vehicles. The company's early products were similar to the more basic automobiles that were sweeping across Europe. But Scania's founders had a clear vision: they wanted to create a product that would revolutionize how goods were transported across the country. They quickly expanded their focus to include trucks, realizing that the future of Swedish industry would depend on reliable vehicles capable of handling long distances and heavy loads.

Though both companies were located in different parts of Sweden, their paths were destined to cross. Both Vabis and Scania faced similar challenges and recognized the importance of innovation in transportation. As they grew, both companies began to understand that collaboration, rather than competition, might offer the best route to success. This understanding eventually led to one of the most significant moments in the history of the Swedish automotive industry.

The Merging of Vabis and Scania to Form Scania-Vabis

In 1911, Vabis and Scania decided to join forces, merging their operations to form Scania-Vabis. The newly formed company was a response to the growing demand for better and more reliable vehicles. By combining the resources and expertise of two strong players in the Swedish automotive industry, Scania-Vabis aimed to create a unified force capable of competing on both national and international levels.

The merger was an ambitious move, bringing together Vabis's experience with building heavy-duty vehicles for industrial use and Scania's innovative design approach to trucks and automobiles. The partnership allowed the company to streamline its operations, consolidate research and development efforts, and increase its production capacity. The newly formed Scania-Vabis quickly established itself as a force to be reckoned with in the automotive world.

The company's first product after the merger, a line of trucks and buses, was a remarkable achievement for the time. They were powerful and durable, built to handle the demanding conditions of Swedish roads and weather. Their efficiency and reliability quickly earned them a reputation, not just in Sweden but across Europe, as a preferred choice for industries that required heavy transportation.

The merger also marked the beginning of Scania-Vabis's international expansion. Recognizing that the global market was on the verge of massive growth, the company started to explore new opportunities outside Sweden. This forward-thinking mindset helped Scania-Vabis set the stage for the company's future dominance in the heavy transport sector.

The Vision of its Founders and Early Product Innovations

The vision of the founders of Scania-Vabis was always one of innovation and practicality. They understood that the future of transportation lay in vehicles that were not only efficient but also capable of handling diverse and challenging conditions. From the start, Scania-Vabis focused on engineering vehicles that could perform reliably across long distances, even in harsh climates. This meant designing trucks and buses that were powerful yet economical, and that could be maintained and repaired easily in remote areas.

One of the company's first innovations was the development of a truck with an advanced engine design that allowed for better fuel efficiency and greater hauling capacity. The early models were designed to meet the specific needs of Swedish industries, such as forestry, mining, and construction, which required vehicles capable of hauling heavy loads over long distances in difficult conditions. These early innovations laid the groundwork for Scania-Vabis's reputation for producing trucks that were not only strong but also efficient and economical to operate.

The company's early trucks and buses were well-suited to Sweden's needs, but they were also designed with an eye on future growth. Scania-Vabis quickly embraced the potential of mass production, seeking to develop vehicles that could be made more efficiently and at a lower cost. This focus on efficiency and quality would become a defining feature of Scania-Vabis, distinguishing it from other companies in the industry.

Through their vision and innovation, the founders of Scania and Vabis transformed the Swedish transport landscape, setting the stage for a company that would become one of the world's leading manufacturers of heavy trucks. The merger of Vabis and Scania in 1911 marked the beginning of a new era in Swedish transportation, one that would see the company expand its influence around the globe and continue to push the boundaries of automotive technology.

Chapter 2: Scania-Vabis Takes the Lead

As the dust settled after the merger of Scania and Vabis in 1911, the newly formed company began its journey of growth and innovation. Scania-Vabis, with its combined expertise and resources, quickly became a dominant force in Sweden's burgeoning automotive industry. The company's early successes were driven by a relentless pursuit of better engineering and the vision of a transportation future that would expand beyond Sweden's borders. The period between the two World Wars would see Scania-Vabis emerge as a leader in the European market, a company capable of meeting the needs of a rapidly changing world.

The First Heavy-Duty Trucks and Buses

The first significant product from Scania-Vabis after the merger was the development of heavy-duty trucks and buses. With a rapidly expanding industrial sector in Sweden and increasing demand for reliable transportation, Scania-Vabis turned its focus to the design of vehicles that could handle larger loads and longer distances. These new trucks were a breakthrough, boasting powerful engines and rugged designs tailored for demanding tasks in industries like mining, forestry, and construction.

The first of these heavy-duty trucks, introduced in the early 1920s, were equipped with advanced engineering features that set them apart from many competitors. Scania-Vabis was one of the first companies to design trucks that were capable of hauling substantial loads over long distances without compromising on engine performance or fuel efficiency. These trucks quickly gained popularity among Swedish businesses that needed reliable vehicles to transport goods across the country's diverse landscape.

In addition to trucks, Scania-Vabis also expanded into the bus market. The company's buses were designed with similar engineering principles, focusing on durability and reliability. With the rapid expansion of cities and towns across Sweden and Europe, public transport was becoming increasingly important. Scania-Vabis recognized this trend early and set about designing buses that could carry large numbers of passengers across long distances, even in challenging conditions.

The development of heavy-duty trucks and buses allowed Scania-Vabis to solidify its position as an industrial leader in Sweden. However, the company had greater aspirations. It was clear that the road ahead lay in expanding beyond the borders of its home country, entering new international markets where its products could make a significant impact.

Expansion Beyond Swedish Borders

As Scania-Vabis strengthened its product offerings and solidified its position in the Swedish market, the company's leadership began to look outward. The potential for growth lay beyond Sweden's borders, particularly in neighboring countries like Denmark, Norway, and Finland, where the demand for reliable trucks and buses was on the rise. But Scania-Vabis was also eyeing even broader horizons—markets that could offer new opportunities in a rapidly industrializing Europe.

The company's first foray into international markets came in the 1920s, when it began to establish a presence in Germany, the Netherlands, and France. These countries, with their burgeoning economies and rapidly growing transportation needs, were prime targets for Scania-Vabis's vehicles. The company's ability to offer durable, efficient trucks and buses that could operate in a variety of environmental conditions made them an appealing choice for industries across Europe.

By the early 1930s, Scania-Vabis had become a well-known name in the European truck and bus market. The company had expanded its manufacturing capacity to meet the increasing demand for its products, and its trucks were now seen as a symbol of quality and dependability in the transportation sector. As a result, Scania-Vabis began to secure important contracts with large industrial players across Europe, helping it establish itself as a key player in the heavy-duty vehicle market.

Scania-Vabis's expansion was not limited to sales alone. The company also began to build relationships with other manufacturers and suppliers, ensuring that its products could be serviced and maintained in international markets. This network of partnerships laid the foundation for Scania-Vabis's long-term growth, allowing the company to maintain a strong presence in foreign markets and to continue innovating its products to meet the needs of global customers.

The Impact of Scania-Vabis in Europe During the 1920s and 1930s

Throughout the 1920s and 1930s, Scania-Vabis's impact on Europe was undeniable. The company's trucks and buses played an essential role in the development of Europe's transportation infrastructure, helping to facilitate the movement of goods and people across the continent. As industries grew and economies expanded, the demand for transportation solutions became increasingly important. Scania-Vabis, with its fleet of rugged and reliable vehicles, quickly positioned itself as one of the leading manufacturers in the European market.

In Germany and the Netherlands, where infrastructure projects were booming, Scania-Vabis's heavy-duty trucks became indispensable for industries like construction and mining. The company's vehicles were known for their ability to carry heavy loads over long distances, even in challenging weather conditions. Whether hauling raw materials across Europe's rugged terrain or transporting workers to distant job sites, Scania-Vabis trucks proved themselves to be reliable partners in the development of Europe's modern industrial landscape.

In France, Scania-Vabis made a significant impact in the passenger transport sector with its line of buses. As cities grew and populations expanded, the demand for efficient, large-capacity buses increased. Scania-Vabis's buses were well-suited to meet this demand, providing reliable and comfortable transport for thousands of passengers daily. The company's reputation for quality engineering and dependability made it a preferred choice for bus operators across Europe.

By the mid-1930s, Scania-Vabis had become synonymous with high-quality, durable vehicles capable of meeting the evolving needs of European industry. The company's ability to adapt its products to the specific requirements of different markets—whether it was designing specialized trucks for long-distance hauling or creating buses for urban transport—allowed it to expand its influence throughout the continent.

Key Personnel and Leadership Shaping the Company's Direction

While the success of Scania-Vabis in the 1920s and 1930s can be attributed to the strength of its products, it was also the vision and leadership of key individuals within the company that helped propel it to new heights. The leadership at Scania-Vabis during this period understood that innovation was key to success, and they were committed to building a company that could thrive not just in Sweden but across the globe.

One of the most influential figures in Scania-Vabis's early growth was **Jörgen W. Håkansson**, who served as managing director during the company's formative years. Håkansson played a critical role in the merger of Scania and Vabis, bringing together the strengths of both companies and steering them toward a shared vision. His leadership and focus on quality control helped establish the company as a leader in Swedish manufacturing, and his emphasis on expanding into new markets laid the foundation for Scania-Vabis's eventual international success.

Another key figure in the company's early success was **Sven A. Sörensen**, who served as the chief engineer and was instrumental in the development of Scania-Vabis's heavy-duty trucks and buses. Sörensen's engineering expertise helped shape the company's vehicles into reliable, powerful machines that could handle the tough conditions of both industrial and commercial use. His forward-thinking approach to vehicle design set Scania-Vabis apart from many of its competitors.

With Håkansson's strategic vision and Sörensen's engineering brilliance, Scania-Vabis was poised for further growth. The company's leadership was not only focused on making high-quality products but also on creating an organization that could continue to innovate, adapt, and grow in a rapidly changing global market. The combination of solid leadership and a strong commitment to product excellence helped Scania-Vabis firmly establish its place as a leader in the European automotive industry during the 1920s and 1930s.

Chapter 3: Scania in the Pre-War Years

The years leading up to World War I were a time of transformation for Scania-Vabis. The company's products, which had already made a name for themselves in Sweden and parts of Europe, were becoming more refined and capable, meeting the demands of an expanding industrial world. However, the outbreak of the Great War in 1914 would have a profound impact on production, forcing the company to shift its focus and adapt to the global conflict. Despite the challenges, the pre-war years were crucial in shaping Scania-Vabis into the formidable company it would become in the interwar period and beyond.

World War I and Its Impact on Production

The outbreak of World War I in 1914 sent shockwaves through Europe and the global economy. For many industries, the war created a significant disruption, as resources were redirected to support the war effort. Scania-Vabis, like many other manufacturers, was forced to adapt its operations to meet the urgent needs of the military. The company's factories, originally focused on producing trucks and buses for civilian industries, now found themselves tasked with producing vehicles and equipment for the war.

Scania-Vabis made significant contributions to the war effort, particularly through the production of military vehicles. The company was already known for its heavy-duty trucks, which made it well-suited to adapt its technology for military use. In addition to standard trucks, Scania-Vabis also developed specialized vehicles, such as field ambulances and military transports, designed to carry troops and supplies across the rough terrain of war zones. These vehicles were critical in maintaining the movement of goods and personnel, especially on the front lines where traditional transportation was often inadequate.

Despite the strain of wartime production, Scania-Vabis managed to keep its core business intact. The company maintained its focus on the development of heavy-duty vehicles, even as it diverted resources to the war effort. By the time the war ended in 1918, Scania-Vabis had not only contributed to the military but also gained invaluable experience in the production of specialized vehicles that would influence the design of its future models.

The war had a lasting effect on Scania-Vabis's operations. While the global economy was in a state of turmoil following the war, Scania-Vabis emerged from the conflict with an expanded knowledge base and a new set of capabilities. The company had built a reputation for durability and innovation, and it was now poised to capitalize on this reputation in the years to come.

Early Developments in Engine Technology

Even before the war, Scania-Vabis had been at the forefront of engine development. The company's early trucks were powered by simple, reliable engines that could handle the heavy demands of long-distance hauling. However, as the demand for more powerful and fuel-efficient vehicles grew, Scania-Vabis was quick to invest in the development of advanced engine technologies.

One of the company's key innovations during this period was the development of the first in-house built engines for its vehicles. Previously, Scania-Vabis had relied on third-party suppliers for its engines, but with the growing complexity of truck design, it became clear that building its own engines would give the company greater control over performance and reliability. In 1912, Scania-Vabis introduced its first in-house engine, a four-cylinder unit that was notable for its strength and durability. This engine became the cornerstone of Scania-Vabis's early trucks, powering vehicles that were capable of carrying heavier loads over longer distances.

In addition to its in-house engine development, Scania-Vabis also focused on refining its designs to improve fuel efficiency and reduce wear and tear on vehicles. As fuel costs increased and the demand for more cost-effective transportation grew, Scania-Vabis's engineers worked to design engines that offered better fuel economy without sacrificing power or performance. These early efforts would lay the foundation for the company's future innovations in engine technology, which would continue to evolve throughout the 20th century.

The company also made strides in the area of safety, introducing new features to its vehicles that would become industry standards in the years to come. This commitment to technological advancement and attention to detail in engine design helped Scania-Vabis establish a reputation for quality and reliability that would become synonymous with the brand.

The Company's Expansion into International Markets and Its Global Reach

While Scania-Vabis had already established a strong presence in Sweden and parts of Europe, the years following World War I marked a new chapter in the company's global expansion. The war had created a challenging economic environment, but it also opened new opportunities for businesses that could adapt to a rapidly changing world. Scania-Vabis, with its established reputation for producing durable and reliable trucks, was well-positioned to take advantage of the post-war recovery.

In the early 1920s, Scania-Vabis began to focus its efforts on expanding beyond Europe. The company started looking at markets in North America, where the demand for heavy-duty trucks and commercial vehicles was on the rise. The United States, in particular, represented a significant opportunity. With its rapidly growing economy and vast road networks, the American market was a promising destination for Scania-Vabis's products.

By the mid-1920s, Scania-Vabis had established a foothold in the United States, where its trucks were well-received by American businesses looking for reliable vehicles to transport goods across the country's vast distances. The company's ability to offer rugged, heavy-duty trucks made it a strong competitor in the American market, which was becoming increasingly competitive with the rise of domestic manufacturers.

At the same time, Scania-Vabis continued to expand its reach across Europe, securing contracts in countries like Germany, France, and the United Kingdom. The company's vehicles were used in a wide range of industries, from construction to agriculture, and its reputation for quality continued to grow. By the end of the 1920s, Scania-Vabis had established itself as one of the leading manufacturers of heavy-duty trucks and buses in Europe, with a growing presence in international markets as well.

This period of expansion set the stage for Scania-Vabis's continued growth in the years leading up to World War II. The company had successfully navigated the challenges of the war and post-war periods, and its ability to adapt to new markets and industries positioned it for long-term success on the global stage.

Notable Pre-War Vehicle Models

Scania-Vabis's success in the early 20th century was built on a series of innovative and reliable vehicle models that helped shape the company's identity. The company's trucks, in particular, became known for their ruggedness and ability to handle heavy loads, even in the most challenging conditions.

One of the most notable early models was the **Scania-Vabis L10**, introduced in 1912. The L10 was a heavy-duty truck designed for long-distance hauling and industrial use. Powered by a four-cylinder engine, it was capable of carrying heavy loads over long distances, making it an ideal choice for industries like mining and forestry. The L10 became one of Scania-Vabis's most popular models, helping to establish the company as a leader in the heavy-duty truck market.

In the years that followed, Scania-Vabis continued to refine its truck designs, introducing more powerful engines and improved features. The **Scania-Vabis 2-axle truck**, introduced in the early 1920s, was another important milestone. This model featured a more powerful engine and better suspension, making it even more reliable and capable of handling heavy loads over rough terrain. The company's buses, too, became known for their comfort and durability, with models like the **Scania-Vabis K4** bus, introduced in 1925, helping to solidify the company's reputation in the passenger transport market.

These early models were just the beginning of Scania-Vabis's journey toward becoming a global leader in the heavy-duty vehicle market. Through their innovation and commitment to quality, Scania-Vabis was laying the groundwork for the company's future successes, both in Sweden and on the international stage.

Chapter 4: World War II and Post-War Rebuilding

The outbreak of World War II in 1939 had a far-reaching impact on industries across the globe, with the Swedish automotive sector not being an exception. The conflict changed the course of history, and for Scania-Vabis, it meant the disruption of civilian production and a forced pivot towards supporting the war effort. During these challenging times, the company not only contributed to the military cause but also emerged from the war stronger, with new technologies and a rejuvenated vision for the future. The post-war years became a period of rebuilding, innovation, and growth as Scania-Vabis turned its attention to the development of new models and technological advancements that would help the company thrive in a changed world.

Scania-Vabis During WWII: Production Challenges and Contributions

At the outset of World War II, Scania-Vabis, like many other manufacturers, was required to adjust its production to support the war effort. Sweden, while officially neutral, found itself at the crossroads of conflict, with both the Allies and Axis powers vying for influence in the region.

As a result, Scania-Vabis was heavily involved in the production of military vehicles and equipment.

With civilian production suspended or limited, Scania-Vabis's manufacturing plants were repurposed for military contracts. The company's primary focus during the war years became the production of trucks, buses, and armored vehicles for the Swedish military and other European nations. These vehicles were crucial for transporting troops, weapons, and supplies to support the war efforts, and Scania-Vabis played a vital role in ensuring that the Swedish military had the vehicles it needed to defend its borders and maintain supply lines.

The company's trucks were often used to carry heavy loads over long distances, navigating the rugged Scandinavian terrain. Scania-Vabis also developed military versions of its existing truck models, making modifications to meet the specific requirements of wartime logistics. These adaptations included the reinforcement of chassis to handle additional weight, as well as the introduction of all-wheel-drive systems to improve the vehicles' off-road capabilities.

While Scania-Vabis's civilian vehicle production was reduced, the war allowed the company to expand its expertise in manufacturing specialized vehicles. The lessons learned from these military contracts would later influence Scania-Vabis's post-war civilian models, which benefited from improved durability and off-road capabilities. Scania-Vabis's reputation for producing rugged and dependable vehicles during the war would serve it well in the years that followed.

Rebuilding and Recovery Efforts After the War

When World War II finally came to an end in 1945, Europe was left in ruins. The immediate post-war years were marked by widespread rebuilding efforts, and for Scania-Vabis, it was a time to shift its focus back to civilian production. However, the company was not simply returning to business as usual. The years of wartime production had given Scania-Vabis valuable experience in adapting to rapidly changing demands, and it was clear that the world of transportation was about to undergo a dramatic shift.

Scania-Vabis, ever forward-thinking, began to reorient itself towards the needs of a post-war world. The demand for heavy-duty trucks and buses had never been higher, as countries across Europe and beyond worked to rebuild their economies. New roads were being constructed, industries were being revitalized, and there was a growing need for more efficient transportation to move goods and people across increasingly larger distances.

Scania-Vabis took advantage of the rebuilding boom by streamlining its production processes, updating its facilities, and refocusing on its core mission of delivering reliable and powerful vehicles. The company began to expand its workforce, bringing in skilled engineers and designers to help develop new models that would meet the needs of the changing world.

Development of New Models and Technological Improvements in the Post-War Era

The post-war era saw Scania-Vabis embark on a new chapter of innovation, building on the lessons learned during the war and introducing vehicles that would set the company apart from its competitors. One of the first major milestones in this period was the development of the **Scania-Vabis L56**, a heavy-duty truck introduced in 1949. The L56 represented a major leap forward in truck design, with improvements in both engine performance and overall reliability. It featured a larger, more powerful engine, better fuel efficiency, and an improved chassis that made it more durable on difficult roads.

In addition to the L56, Scania-Vabis also began to invest heavily in the development of new bus models. As cities expanded and populations grew, the need for reliable public transport was more pressing than ever. Scania-Vabis responded with buses that were not only durable and efficient but also designed with passenger comfort in mind. The **Scania-Vabis B52** bus, introduced in the early 1950s, became a staple in cities across Europe, offering modern features such as improved seating arrangements, better ventilation, and enhanced safety features.

During this period, Scania-Vabis also began to focus more on developing multi-axle trucks and trailers. These vehicles were designed to carry heavier loads and were crucial for industries like construction, mining, and agriculture, which were at the heart of post-war rebuilding efforts. The ability to transport larger amounts of goods efficiently became a key competitive advantage, and Scania-Vabis's new multi-axle vehicles quickly became sought after by businesses looking to move materials across long distances.

Technologically, the post-war years marked a period of rapid innovation for Scania-Vabis. The company began to experiment with more advanced materials, incorporating lighter and more durable metals into its vehicles, and introduced advanced transmission systems that improved both performance and fuel economy. The engines themselves were refined to be more powerful, yet more fuel-efficient, helping to reduce operating costs for customers. Scania-Vabis also developed new safety features, such as improved braking systems, which were critical for the heavier and more powerful vehicles being produced at the time.

These innovations were not only driven by a desire to meet market demands but also by Scania-Vabis's commitment to pushing the boundaries of automotive technology. The company's engineers were determined to create vehicles that were not only functional but also ahead of their time in terms of both performance and design.

The Impact of War on the Swedish Automotive Industry

Sweden, unlike many of its European neighbors, had managed to avoid significant physical destruction during the war, owing to its neutral stance. However, the impact of the war on the Swedish automotive industry was still profound. Sweden's strategic position and manufacturing prowess made it an important supplier of goods and services during the conflict, and many companies, including Scania-Vabis, found themselves shifting focus to support the war effort.

The Swedish automotive industry emerged from the war in a unique position. While many countries had to rebuild entire industries from the ground up, Sweden had retained a strong manufacturing base, which allowed companies like Scania-Vabis to ramp up production quickly. However, the war also brought with it increased competition. Companies from other parts of Europe, particularly the United States and Germany, were eager to reassert their dominance in the global automotive market. This competition pushed Scania-Vabis to innovate and improve its products, which would ultimately lead to the rapid growth and success of the company in the post-war years.

The war also influenced the Swedish automotive market in terms of design and engineering priorities. The necessity for vehicles capable of operating in challenging and diverse environments during the war had led to a focus on durability, versatility, and reliability—qualities that continued to shape the Swedish automotive industry in the years that followed. Scania-Vabis, in particular, benefited from these lessons, as the vehicles it produced in the post-war era were built to withstand harsh conditions and demanding workloads.

In conclusion, the years during and immediately after World War II were a time of profound change and growth for Scania-Vabis. The company's wartime contributions not only helped it gain experience in specialized vehicle production but also reinforced its reputation for producing durable and reliable vehicles. As the world rebuilt itself after the war, Scania-Vabis emerged stronger, armed with new technologies and a renewed focus on innovation. The company's commitment to quality and performance would continue to guide its development in the decades to come, ensuring its place as a leader in the global transport industry.

Chapter 5: The 1950s and 1960s: A New Age of Heavy-Duty Trucks

The post-war era saw significant transformations in the global transport sector. As industries rebuilt and the demand for goods surged, the need for more efficient, powerful, and reliable vehicles became paramount. Scania-Vabis, having navigated the challenges of the war years and the post-war recovery, was ready to capitalize on this newfound demand. The 1950s and 1960s marked a pivotal moment in the company's history, as it entered a new age of heavy-duty trucks. During this time, Scania-Vabis not only innovated its product lineup but also expanded into new international markets, setting the stage for its global success.

The Shift Towards Larger, More Powerful Trucks

By the early 1950s, the global economy was on the rise. With industries booming and urban populations growing, the need for more efficient transportation was greater than ever. Scania-Vabis recognized this shift and began adapting its product range to meet the growing demand for larger, more powerful trucks capable of carrying heavier loads over longer distances.

As the world became more interconnected, the need for trucks that could handle the increasing volume of goods being moved across countries and continents was clear. In response, Scania-Vabis introduced a new generation of heavy-duty vehicles with larger engines and higher payload capacities. This shift was not just about increasing size and power; it was also about improving the overall design and engineering of the vehicles. Trucks were now expected to perform at higher speeds and under greater stress, all while maintaining fuel efficiency and reducing operational costs.

The 1950s saw the introduction of new truck models that reflected this philosophy. Scania-Vabis focused on designing vehicles that were more robust, more fuel-efficient, and more comfortable for drivers. The larger trucks of this era were built to tackle the increasingly demanding requirements of industries like construction, mining, and long-distance transport. These trucks became symbols of the company's commitment to engineering excellence, with the larger engines and improved powertrains becoming key selling points for businesses that relied on heavy-duty transportation.

Introduction of Key Models and Design Philosophies

Among the most significant introductions during the 1950s and 1960s was the **Scania-Vabis L75**, launched in 1954. The L75 was a breakthrough in the heavy-duty truck market, combining power, durability, and efficiency in a way that had not been seen before. With a more powerful engine, better suspension, and improved aerodynamics, the L75 was designed to carry larger payloads and operate in more demanding environments. The L75 was a commercial success, both in Sweden and abroad, and became one of Scania-Vabis's defining models during this period.

Building on the success of the L75, Scania-Vabis introduced the **Scania-Vabis L80** in the late 1950s. The L80 was a further refinement of the L75, featuring even more powerful engines, improved braking systems, and an updated cab design. The vehicle's increased payload capacity and operational efficiency made it an ideal choice for industries that required heavy-duty transportation for long distances. The L80 was not just a truck; it was a symbol of the company's engineering prowess, embodying the values of strength, reliability, and innovation that Scania-Vabis was becoming known for.

During this time, Scania-Vabis also focused on the driver experience. Understanding that drivers spent long hours on the road, the company made improvements to the comfort and ergonomics of its vehicles. The cabs of the L75 and L80 were designed with more spacious interiors, better visibility, and improved driving controls, making the driving experience more comfortable and safer. These changes helped position Scania-Vabis as a leader in the heavy-duty truck market, with a growing reputation for producing not just powerful vehicles but also ones that prioritized driver well-being.

Expansion into New International Markets, Particularly in Europe and the Americas

As Scania-Vabis continued to innovate its product range, it also expanded its reach beyond Sweden. The company had already established a foothold in neighboring European markets in the 1920s and 1930s, but the post-war economic boom provided a new opportunity for broader international expansion. The 1950s and 1960s marked a period of aggressive growth as Scania-Vabis sought to extend its presence in Europe and even across the Atlantic to the Americas.

The expansion into the European market was a natural step for Scania-Vabis, as many of the countries in Europe were in the midst of rebuilding their economies and required reliable heavy-duty vehicles to support industrial growth. Scania-Vabis capitalized on this by offering trucks that were durable enough to handle the tough conditions of European roads and efficient enough to meet the growing demands of industries in countries like Germany, France, and the Netherlands.

However, it was Scania-Vabis's expansion into the American market that truly set the company on the path to becoming a global leader. By the late 1950s, the company had begun to export its trucks to the United States, a market that was dominated by American manufacturers. Breaking into this market was no small feat, but Scania-Vabis's reputation for quality, durability, and engineering excellence gave it a competitive edge. The company's trucks were particularly well-received in the United States for their ability to carry large loads over long distances, which made them ideal for the vast highway systems that spanned the country.

Scania-Vabis also made significant strides in Latin America, where the demand for heavy-duty vehicles was growing in response to expanding industries and infrastructure projects. The company's ability to provide reliable and efficient trucks that could withstand the region's varied climates and rugged terrains helped it establish a strong foothold in this important market.

By the end of the 1960s, Scania-Vabis had firmly established itself as a global player in the heavy-duty truck market. The company's ability to adapt to diverse international markets, coupled with its strong reputation for quality and innovation, made it one of the leading manufacturers of commercial vehicles in the world.

Technological Milestones During This Era

The 1950s and 1960s were a time of significant technological advancement for Scania-Vabis. The company's engineering team continued to push the boundaries of what was possible, introducing new technologies that would shape the future of heavy-duty trucks for decades to come.

One of the most significant technological milestones during this period was the development of Scania-Vabis's **V8 engine**, introduced in the early 1960s. The V8 engine was a game-changer, offering superior power and performance compared to the standard six-cylinder engines used in trucks at the time. The V8 engine allowed Scania-Vabis to produce trucks that could handle even heavier loads and travel longer distances, making them ideal for long-haul trucking operations. The introduction of the V8 engine also set Scania-Vabis apart from many of its competitors, as it demonstrated the company's commitment to pushing the limits of engine technology.

In addition to the V8 engine, Scania-Vabis introduced several other innovations in the areas of fuel efficiency and safety. The company refined its suspension systems to improve ride comfort and handling, particularly for long-distance drivers. Scania-Vabis also made significant improvements in its braking systems, introducing air-assisted brakes that were more reliable and effective than traditional systems. These technological advancements made Scania-Vabis trucks safer to operate and more efficient to run, helping the company maintain its competitive edge in an increasingly globalized market.

The 1950s and 1960s were formative years for Scania-Vabis, marking the transition from a regional manufacturer of heavy-duty vehicles to a global leader in the industry. With its new models, expanding international presence, and groundbreaking technological innovations, Scania-Vabis was well-positioned to thrive in the rapidly changing landscape of the 20th century transport sector. The company's commitment to quality, innovation, and customer satisfaction would continue to guide its growth, ensuring its place at the forefront of the global trucking industry for decades to come.

Chapter 6: Scania in the 1970s: Innovation and Global Expansion

The 1970s were a decade of profound change for the global transport industry. Economic fluctuations, geopolitical shifts, and technological advances created a landscape where manufacturers needed to adapt quickly to remain competitive. Scania, which had already established itself as a leader in the heavy-duty truck market, entered the decade with a renewed focus on innovation and expansion. During this period, the company introduced new engine technologies and truck models, faced the challenges posed by the oil crisis, and undertook strategic acquisitions and mergers that would lay the foundation for its long-term global success. The decade also marked a period of significant growth, as Scania sought to strengthen its presence not just in Europe but in markets across the world.

New Engine Technologies and Truck Models

The 1970s saw some of Scania's most important innovations in engine technology, as the company sought to maintain its competitive edge in an increasingly crowded global market. The demand for more powerful, efficient, and environmentally friendly engines was growing, and Scania responded by developing new powertrains that would set the standard for the heavy-duty truck industry.

One of the most significant advancements in this period was the introduction of the **Scania V8 engine** in 1971. Building on the success of the earlier V8 engine that had been introduced in the 1960s, Scania's 1970s version featured even greater power and efficiency. With its larger displacement and more advanced engineering, the V8 engine was able to produce significantly more horsepower, making it one of the most powerful engines in the heavy-duty truck market at the time. The V8 engine quickly became a hallmark of Scania's vehicle lineup, used in a variety of applications, from long-haul trucking to construction and mining.

The V8 engine was part of a broader shift in Scania's truck design philosophy. The company began focusing more on developing trucks that were not only powerful but also fuel-efficient, able to meet the growing demands for longer trips with fewer refueling stops. Scania's engineers continued to refine the performance of the V8, improving fuel consumption and operational costs, which allowed the company's trucks to remain highly competitive in markets where operating costs were becoming an increasing concern.

In addition to the V8, Scania introduced the **Scania 2 Series** in the early 1970s. The 2 Series was a complete redesign of the company's truck lineup, featuring updated styling, more powerful engines, and better aerodynamics. The series was aimed at long-haul operators who needed trucks that could handle greater distances and larger payloads while maintaining driver comfort and safety. The 2 Series also featured Scania's new **Opticruise** transmission system, a pioneering development that allowed for smoother gear shifts and reduced driver fatigue. This was a crucial step in Scania's goal of improving the driver experience while increasing operational efficiency.

The 2 Series was an immediate success, earning praise for its power, reliability, and innovative features. The introduction of these advanced truck models positioned Scania as one of the most forward-thinking manufacturers in the heavy-duty segment, capable of addressing the evolving needs of customers in a changing market.

The Impact of the Oil Crisis on Scania's Design and Production Strategy

The global oil crisis of the 1970s had a profound impact on the automotive industry, especially for manufacturers of heavy-duty trucks. In 1973, oil prices skyrocketed as a result of geopolitical tensions in the Middle East, leading to fuel shortages and inflation across much of the world. For Scania, this created a challenging environment. The demand for more fuel-efficient trucks grew significantly, as companies sought to reduce operating costs in the face of higher fuel prices.

Scania's response to the oil crisis was to focus even more heavily on improving the fuel efficiency of its trucks. The company invested in new engine technologies that were not only more powerful but also more fuel-efficient, which allowed Scania's vehicles to remain cost-effective despite the rising price of fuel. This was achieved through a combination of engine refinements, improved aerodynamics, and better weight distribution across the trucks.

One key innovation during this time was the **Scania Super** engine, introduced in 1976. The Super engine was designed to deliver more power with less fuel, making it ideal for the long-haul operators who were feeling the pressure of rising fuel costs. This engine, paired with the company's continued focus on lightweight truck designs, helped Scania maintain its reputation for offering efficient, reliable, and cost-effective vehicles.

Scania also worked on improving the overall efficiency of its truck lineup by rethinking the way trucks were built. The company focused on reducing the weight of its vehicles, which improved fuel efficiency without sacrificing strength or durability. By introducing new materials and streamlining production processes, Scania was able to offer trucks that were not only more fuel-efficient but also more competitive in an increasingly price-sensitive market.

The oil crisis of the 1970s had a lasting effect on Scania's design and production strategy, pushing the company to focus on long-term sustainability through fuel efficiency and operational cost reduction. These efforts helped Scania weather the storm of rising fuel prices and position itself for future growth as the global market began to recover.

Strategic Acquisitions and Mergers in the 1970s

During the 1970s, Scania-Vabis also took steps to strengthen its position through strategic acquisitions and mergers. One of the most important moves came in 1970, when Scania-Vabis merged with **Värmland** to form a more robust, competitive entity. This merger allowed Scania to increase its production capacity and broaden its product offerings, making the company even more competitive in global markets.

The merger with Värmland allowed Scania-Vabis to streamline operations, reduce costs, and expand its reach into new markets. It also provided the company with access to new manufacturing technologies, which helped Scania maintain its leadership in heavy-duty vehicle design. The combined company was able to take advantage of economies of scale, enabling it to produce trucks and buses more efficiently and at a lower cost, which proved to be a significant advantage in a competitive market.

In addition to mergers, Scania also expanded its dealer networks and strengthened partnerships with other manufacturers. By aligning itself with global distributors and suppliers, Scania was able to increase its market share and build a stronger presence in key international markets. The company also began to form strategic alliances with other vehicle manufacturers, allowing for shared technological advancements and mutual growth.

Strengthening Scania's Global Presence

The 1970s marked a period of aggressive expansion for Scania, as the company sought to strengthen its position in international markets, particularly in Europe and the Americas. Scania had already established a solid foundation in Europe, but the 1970s saw a deliberate push into new markets that were growing in importance.

In North America, Scania made significant strides by introducing its trucks into the United States and Canada. The North American market was competitive, dominated by well-established American manufacturers, but Scania's reputation for reliability, fuel efficiency, and performance gave it a competitive edge. The company quickly garnered interest from fleet operators, especially those involved in long-haul trucking, who appreciated Scania's fuel-efficient designs during the oil crisis.

In Latin America, Scania began to see greater demand for its trucks, driven by industrial expansion in countries like

Brazil and Argentina. The rugged terrain and long-distance transportation needs of the region made Scania's vehicles an ideal choice. The company also built strong relationships with local distributors and service centers, ensuring that its trucks were well-supported in these rapidly developing markets.

By the end of the decade, Scania's trucks were on the roads of nearly every continent, from Europe and North America to South America, Africa, and Asia. The company's expanding global presence was a testament to its commitment to innovation and quality, as well as its ability to adapt to the needs of diverse international markets. The 1970s laid the groundwork for Scania's continued growth and cemented its status as one of the world's leading manufacturers of heavy-duty trucks.

In conclusion, the 1970s were a transformative decade for Scania. The company navigated the challenges of the oil crisis, introduced groundbreaking new technologies, and expanded its global footprint. Strategic mergers and acquisitions helped solidify Scania's position in the market, and its focus on innovation ensured that it would remain a competitive force in the heavy-duty truck industry for decades to come. Scania emerged from the

1970s stronger, more resilient, and more focused on the future than ever before.

Chapter 7: The 1980s: A Decade of Transformation

The 1980s were a defining decade for Scania, one that marked a period of dramatic transformation and innovation. As the global transport industry evolved, Scania found itself at the forefront of this change, with new truck models, technological advancements, and expansion into new markets. This was a time when the company not only consolidated its position as a leader in heavy-duty trucks but also adapted to meet the challenges of a rapidly changing market. From the launch of the **Scania 2 Series** to the push for greater fuel efficiency, the 1980s were a period of reinvention for Scania, laying the foundation for its future growth and success.

The Launch of the Scania 2 Series and the Revolution in Truck Design

One of the most significant events of the 1980s for Scania was the launch of the **Scania 2 Series**. Introduced in 1980, the 2 Series represented a complete redesign of the company's truck lineup, and it was a revolutionary step in truck design. The 2 Series was not just an evolution of Scania's previous models—it was a statement of intent, showcasing the company's commitment to innovation, driver comfort, and performance.

The 2 Series was designed with a focus on three key elements: power, reliability, and driver comfort. The truck's chassis was more robust, and its engine options were more powerful and efficient than ever before. Scania introduced new, larger engines to provide more torque and better acceleration, making the trucks even more capable of handling heavy loads over long distances. The range of models included everything from short-haul trucks to long-distance haulers, each designed to meet the specific needs of various industries.

One of the key innovations of the 2 Series was its **cab design**. Scania redefined the truck cabin, making it more spacious, comfortable, and functional. This was a significant shift in the industry, where driver comfort had often been an afterthought. The 2 Series featured an improved driving position, better visibility, and better ergonomics, making it easier for drivers to spend long hours on the road. The improved interior also included more advanced features, such as better insulation for noise and temperature control, contributing to a more comfortable driving experience.

The 2 Series marked a new chapter for Scania, positioning the company as not just a producer of powerful and reliable trucks, but as a manufacturer that truly cared about the needs of the driver. The launch of this series was a defining moment in Scania's history, as it helped establish the company's reputation for innovation and quality, both in terms of performance and driver experience.

The Company's Response to the Growing Demand for Fuel Efficiency

As the 1980s progressed, the global transportation industry continued to face the challenge of rising fuel prices and increased environmental awareness. Fuel efficiency had become a critical factor for fleet operators, and Scania was quick to recognize the need for more fuel-efficient trucks. The company responded by focusing on improving engine technology and vehicle design to help operators reduce their fuel consumption.

Scania introduced several measures to address this challenge. The company's engineers worked tirelessly to improve engine efficiency, incorporating new technologies that allowed for better combustion and lower fuel consumption. This included the development of **turbocharged engines**, which provided more power with less fuel. The turbocharged engine allowed Scania trucks to carry heavier loads while still maintaining impressive fuel economy, making them highly attractive to long-haul operators.

In addition to improving engine performance, Scania also focused on the overall design of its trucks. The aerodynamics of vehicles became a key consideration in reducing fuel consumption, especially as trucks began to travel longer distances across highways and motorways. Scania invested in developing trucks with better aerodynamic profiles, reducing drag and improving fuel efficiency on the open road. The introduction of more streamlined cabs and smoother chassis designs helped improve fuel economy, contributing to the company's growing reputation as a manufacturer of fuel-efficient, long-haul trucks.

Scania's focus on fuel efficiency also extended to its heavy-duty vehicles. As industries around the world faced pressure to reduce operating costs, Scania's fuel-efficient trucks became highly sought after, especially in markets where long-distance hauls and high payloads were the norm. By the mid-1980s, Scania had positioned itself as one of the industry leaders in fuel-efficient transportation, giving it a competitive advantage as the demand for more sustainable solutions grew.

Technological Advancements in Transmission, Engines, and Aerodynamics

The 1980s were marked by significant technological advancements in Scania's product lineup, particularly in the areas of transmission, engines, and aerodynamics. As the demands of the transport industry grew more complex, Scania responded by introducing innovations that would enhance both performance and efficiency.

One of the key developments of the decade was the introduction of **Scania's 8-speed transmission system**. The new transmission system allowed for smoother gear changes and improved fuel efficiency, particularly when driving on highways and long-distance routes. The system made it easier for drivers to handle a wide range of road conditions, while also reducing wear and tear on the engine. This advancement was part of Scania's broader strategy to improve the drivability of its vehicles, ensuring that long-haul drivers could spend more time on the road without experiencing fatigue or discomfort.

In terms of engine technology, Scania continued to innovate with more powerful and efficient engines, building on the success of its turbocharged and V8 engines. During this period, Scania introduced new engines with lower emissions, helping to meet the increasing environmental regulations that were becoming more prevalent across Europe and beyond. The company's commitment to developing cleaner, more efficient engines allowed it to stay ahead of regulatory changes and continue to meet the needs of a rapidly evolving market.

Aerodynamics, too, became a key focus for Scania in the 1980s. With fuel efficiency becoming more critical, the company began to experiment with truck designs that were not only more streamlined but also better able to reduce wind resistance. The introduction of more rounded cab designs and improved front grills helped reduce drag, making the trucks more fuel-efficient and easier to drive at higher speeds. These aerodynamic improvements were crucial for long-distance trucking, where fuel savings could translate into significant operational cost reductions.

Scania's Entry into New Markets, Including North America

In the 1980s, Scania set its sights on expanding its global reach, particularly into markets where it had not yet established a strong presence. North America, with its vast road networks and well-established trucking culture, became a key target for Scania's growth.

Entering the North American market was a significant challenge for Scania. The region was dominated by large American truck manufacturers, and many fleet operators were loyal to domestic brands. However, Scania's reputation for reliability, fuel efficiency, and driver comfort made it an attractive alternative for fleet operators seeking to improve their operations.

The company made a concerted effort to design trucks specifically suited for the North American market, where long-haul trucking was the norm and the roads were often more demanding. Scania introduced models with larger engines and more powerful transmissions to cater to the region's trucking needs. The trucks were also adapted to meet local regulatory requirements, including emissions standards and weight restrictions, which were different from those in Europe.

Despite the challenges, Scania's entry into North America was successful. The company built a network of dealers and service centers across the United States and Canada, offering its trucks alongside local brands. Over time, Scania's reputation for high-performance trucks with advanced features helped it establish a foothold in the market, though it would take several years before the company could challenge the dominance of American manufacturers.

In addition to North America, Scania also expanded its presence in other international markets, particularly in Asia and South America. The 1980s saw Scania trucks on roads in countries like Brazil, Argentina, and Japan, as the company sought to capitalize on the growing demand for heavy-duty vehicles in these rapidly developing regions.

Conclusion

The 1980s were a period of dramatic transformation for Scania. The company embraced innovation and responded to the changing demands of the transport industry with new truck designs, technological advancements, and a focus on fuel efficiency. The launch of the Scania 2 Series redefined truck design, while improvements in engine technology, transmission systems, and aerodynamics ensured that Scania's vehicles remained at the cutting edge of the industry. The company's entry into new markets, particularly in North America, helped solidify its position as a global leader in heavy-duty trucks. As the decade came to a close, Scania was more prepared than ever to face the challenges of the next era, with a strong foundation of innovation, customer trust, and global expansion.

Chapter 8: The 1990s: Emissions, Efficiency, and the Modular System

The 1990s were a pivotal decade for Scania, as the company navigated an increasingly complex landscape shaped by environmental concerns, stricter regulations, and a growing need for efficiency in the transport industry. As the world became more focused on sustainability, Scania responded with a series of innovations aimed at reducing emissions, increasing fuel efficiency, and offering greater flexibility to its customers. The introduction of the **Modular System** marked a turning point in the company's approach to truck design, allowing for increased customization and adaptability. The 1990s also saw Scania expanding its truck range to meet the diverse and evolving needs of the global market while continuing its efforts to reduce environmental impact. Through these advancements, Scania solidified its reputation as a leader in both performance and sustainability.

The Introduction of Stricter Emission Standards and Scania's Response

The 1990s marked a significant shift in global environmental awareness. As concerns about air pollution and climate change began to take center stage, governments worldwide introduced stricter regulations on vehicle emissions. Europe, in particular, was at the forefront of these efforts, with the **Euro 1** emissions standard being introduced in 1992. These regulations required manufacturers to reduce the levels of harmful emissions, such as nitrogen oxides (NOx), particulate matter (PM), and hydrocarbons, produced by their vehicles. For Scania, this represented both a challenge and an opportunity to innovate.

Scania responded to the Euro 1 regulations by developing new engine technologies designed to meet the tighter emissions limits while maintaining the performance and reliability the company was known for. The company focused on improving the combustion process within its engines, using advanced technologies such as **electronically controlled fuel injection** and **exhaust gas recirculation (EGR)**. These innovations helped reduce NOx emissions by improving the efficiency of the combustion process, while maintaining the power and torque required for long-haul trucking.

In addition to these technological improvements, Scania also began experimenting with **selective catalytic reduction (SCR)**, a technology that uses a urea-based solution to reduce NOx emissions. This technology would later become a key component of Scania's strategy for meeting the even stricter emissions standards that would follow in the next decade.

As the 1990s progressed, the company continued to invest in research and development to stay ahead of evolving regulations. Scania's ability to produce trucks that met the demands of the new environmental standards while maintaining high performance helped it maintain a competitive edge in the European market, where the Euro 1 standards had set a new benchmark for truck emissions.

Development of the Scania Modular System for Increased Flexibility

One of the most significant innovations of the 1990s was the introduction of the **Scania Modular System**. The modular approach to truck design was developed to offer customers increased flexibility and customization in how they configured their vehicles. Prior to this, truck designs were often rigid, with limited options for customization. Scania's new modular system changed that by allowing customers to select from a variety of components—such as chassis, cabs, axles, and engines—to create a truck that was perfectly suited to their specific needs.

The system was built around a series of standardized modules that could be easily combined to create different types of vehicles. This approach allowed for the efficient production of a wide range of trucks, from light-duty delivery vehicles to heavy-duty long-haul rigs, without compromising on quality or performance. The modular system also made it easier for Scania to adapt its trucks to meet the specific requirements of different markets and industries.

The modular approach offered several advantages. It streamlined the production process, reduced lead times, and lowered costs for both Scania and its customers. For fleet operators, the system provided a level of flexibility that had previously been unattainable. They could now choose the exact specifications they needed, whether it was a more powerful engine, a specific cab design, or an extended chassis to accommodate different types of cargo.

The introduction of the Modular System also marked a shift in Scania's production philosophy. The company was now able to produce vehicles with greater efficiency, reducing waste and improving overall manufacturing processes. At the same time, it allowed Scania to continue its commitment to quality by ensuring that all components were rigorously tested and built to the company's high standards.

Expansion of the Scania Truck Range to Meet Diverse Customer Needs

In the 1990s, Scania's truck range expanded significantly to meet the growing and diverse needs of its global customer base. With industries in Europe, North America, and beyond demanding more specialized vehicles, Scania adapted its offerings to provide solutions for an increasingly varied set of applications.

The company introduced a wider range of engine options, with more powerful engines tailored for specific tasks. This expansion allowed Scania to serve markets that required specialized trucks, such as those in the construction, forestry, and mining sectors. The new models were designed to handle different types of terrain and operate in more demanding environments, whether it was navigating mountain roads, driving through harsh weather conditions, or hauling heavy equipment.

Scania also began to focus more on regional variations, developing models that were tailored to the unique needs of specific markets. In North America, for example, the company introduced trucks with larger engines and more robust chassis to meet the demands of the long-haul industry, where trucks needed to cover vast distances and carry heavier loads. Meanwhile, in Europe, Scania continued to offer a range of models designed for both short-distance urban deliveries and long-haul trucking.

This diversification of Scania's truck range allowed the company to build a broader customer base, from small businesses needing delivery trucks to large fleet operators requiring long-distance vehicles. The ability to offer customized solutions helped Scania stay competitive in an increasingly globalized market, where customers were seeking more specialized and flexible options.

The Focus on Fuel-Efficient Engines and Reducing Environmental Impact

As fuel prices remained volatile and environmental concerns continued to grow, Scania maintained its focus on developing fuel-efficient engines. The company understood that the future of the transport industry would rely not just on performance and reliability but also on reducing the environmental impact of its vehicles.

Scania's engineers worked on developing engines that were more fuel-efficient, with a particular focus on reducing the amount of fuel consumed during long-distance hauls. The introduction of **intercoolers** and **electronic fuel management systems** helped improve the combustion process, ensuring that more of the energy from the fuel was converted into usable power, rather than being wasted as heat.

The company also invested heavily in improving aerodynamics, not just through the design of the truck but also through innovations in the trailers that accompanied them. By making both the truck and the trailer more streamlined, Scania was able to reduce drag and improve fuel efficiency. The introduction of more aerodynamic truck cabs and trailers became a key part of Scania's approach to sustainability, helping customers reduce their fuel costs and their carbon footprint.

By the end of the decade, Scania had positioned itself as one of the leaders in the field of sustainable transportation, with a product lineup that was not only more fuel-efficient but also better for the environment. The company's commitment to reducing emissions and improving fuel efficiency was central to its business strategy as it entered the new millennium, setting the stage for the development of cleaner technologies in the years to come.

Conclusion

The 1990s were a transformative decade for Scania, as the company responded to new challenges and opportunities in the transport industry. The introduction of stricter emission standards pushed Scania to innovate, leading to the development of cleaner, more fuel-efficient engines. The launch of the **Modular System** revolutionized truck design, offering customers greater flexibility and customization than ever before. Scania's expanded truck range allowed the company to serve a more diverse set of industries and markets, while its ongoing focus on fuel efficiency and sustainability solidified its position as a leader in the global transport sector. As the decade drew to a close, Scania's commitment to innovation and environmental responsibility ensured that it was well-positioned to continue its success in the 21st century.

Chapter 9: Entering the 21st Century: Scania's Focus on Sustainability

The turn of the century was a time of rapid transformation for the global transport industry, and Scania's response to these changes would set the stage for the company's continued growth and success. As environmental concerns became a greater priority and the demand for more fuel-efficient and sustainable transportation solutions increased, Scania embraced sustainability as a core component of its strategy. The early 2000s saw the company make significant strides in technological innovation, safety advancements, and fuel efficiency, all while maintaining its reputation for performance and reliability. Scania's commitment to reducing emissions and improving fuel economy was central to its vision, as it continued to evolve in response to the growing need for greener and more efficient transport solutions.

Technological Innovations and Safety Advancements in the 2000s

As the world entered the new millennium, Scania remained at the forefront of technological innovation. The 2000s were marked by a significant leap forward in the development of more advanced truck technologies, including improvements in vehicle safety, driver assistance systems, and on-board connectivity.

Scania focused heavily on developing vehicles equipped with cutting-edge safety features. The introduction of **Electronic Stability Program (ESP)** and **Anti-lock Braking Systems (ABS)** became standard features on many Scania trucks, improving vehicle control and reducing the risk of accidents. Scania's emphasis on safety also led to the development of advanced **driver assistance systems (ADAS)**, which provided real-time data to help drivers avoid potential hazards. These systems included adaptive cruise control, lane-keeping assist, and collision warning systems, all of which contributed to safer roads and lower accident rates.

In addition to safety innovations, Scania also pioneered the integration of **telematics** and **fleet management solutions** into its trucks. By equipping trucks with sensors and GPS technology, Scania allowed fleet operators to monitor real-time performance data, including fuel consumption, maintenance needs, and driver behavior. This connected technology not only improved operational efficiency but also helped reduce fuel consumption and minimize the environmental impact of transportation.

Scania's dedication to technological advancements in safety and efficiency was crucial for meeting the demands of an increasingly regulated and competitive market. The company's investments in these areas helped ensure that it remained a leader in the global truck industry, while also making a significant contribution to improving road safety and environmental sustainability.

The Development of More Efficient Diesel Engines and Hybrid Solutions

Throughout the 2000s, Scania's engine technology continued to evolve, with a strong focus on improving fuel efficiency and reducing emissions. As diesel engines were still the dominant choice for heavy-duty trucks, Scania's engineers worked tirelessly to develop more efficient, low-emission diesel engines that would meet both regulatory requirements and the demands of customers looking to reduce operating costs.

One of the key innovations during this period was the development of **Scania's 9-liter and 13-liter engines**, which offered a significant improvement in fuel efficiency compared to previous models. These new engines incorporated advanced fuel injection systems and turbocharging technologies to improve combustion efficiency, delivering more power with less fuel. This made Scania trucks even more attractive to fleet operators, as they could reduce fuel costs without sacrificing performance.

In addition to improving diesel engines, Scania also began exploring **hybrid solutions** in the early 2000s. The company's goal was to create vehicles that could run on a combination of diesel and electric power, reducing emissions and fuel consumption, particularly in urban environments where short trips and frequent stops contributed to higher fuel use. Scania's hybrid trucks, which combined a traditional diesel engine with an electric motor, provided a more sustainable solution for city-based transport operations.

The development of hybrid solutions was an important step toward Scania's long-term goal of creating a more sustainable future for the transport industry. Although hybrid technology was still in its early stages during the 2000s, Scania's commitment to developing alternative powertrains demonstrated its foresight in addressing the growing demand for cleaner, more efficient vehicles.

Scania's Commitment to Reducing Emissions and Increasing Fuel Economy

As environmental concerns intensified in the 21st century, reducing emissions and improving fuel economy became central priorities for Scania. The company made a significant commitment to addressing these issues, developing strategies to reduce the environmental impact of its trucks while maintaining the high level of performance and reliability that customers expected.

In 2005, Scania launched its **"Green Truck"** initiative, which focused on creating environmentally friendly vehicles with lower emissions and better fuel efficiency. The program aimed to meet the stringent emission standards set by European regulators, particularly the **Euro 3 and Euro 4 emission standards**. These standards required manufacturers to reduce the levels of harmful pollutants such as nitrogen oxides (NOx) and particulate matter (PM) in their engines, prompting Scania to develop new technologies that would meet these regulations.

Scania's response to these challenges was multifaceted. The company introduced **Exhaust Gas Recirculation (EGR)** and **Selective Catalytic Reduction (SCR)** technologies, both of which helped reduce NOx emissions and improve fuel efficiency. EGR works by recirculating a portion of the exhaust gases back into the engine to lower combustion temperatures and reduce NOx formation, while SCR uses a urea-based solution to convert NOx into harmless nitrogen and water vapor.

By incorporating these technologies, Scania's trucks became among the cleanest and most fuel-efficient on the market. The company's dedication to reducing emissions not only helped it meet regulatory standards but also made it an attractive partner for customers looking to reduce their environmental footprint and operating costs.

The Introduction of the Euro 3 and Euro 4 Emission-Compliant Trucks

In line with the evolving regulatory landscape, Scania introduced a series of emission-compliant trucks designed to meet the increasingly stringent European Union emission standards. The **Euro 3** standards, introduced in 2000, required manufacturers to reduce NOx emissions and particulate matter from diesel engines. Scania's response was to implement advanced engine technologies, including the use of EGR and enhanced fuel injection systems, to achieve compliance with the new regulations.

The **Euro 4** standards, which came into effect in 2005, were even stricter, requiring a further reduction in NOx and particulate emissions. To meet these new standards, Scania introduced its next generation of engines, equipped with SCR technology. This allowed Scania to produce trucks that not only met the Euro 4 requirements but also provided better fuel economy, offering fleet operators a more sustainable and cost-effective solution for long-distance transport.

The introduction of Euro 3 and Euro 4 compliant trucks was a significant milestone for Scania, as it demonstrated the company's commitment to environmental sustainability while maintaining its reputation for performance and reliability. By embracing these regulatory changes, Scania not only ensured that its trucks remained competitive in the European market but also set the stage for future innovations in emissions reduction and fuel efficiency.

Scania's Evolving Role in the Global Truck Market

By the 2000s, Scania had firmly established itself as a global leader in the truck manufacturing industry. Its commitment to innovation, sustainability, and performance had earned the company a reputation for producing some of the most reliable, fuel-efficient, and environmentally friendly trucks on the market. As the demand for more sustainable transportation solutions continued to rise, Scania's focus on reducing emissions and improving fuel economy helped solidify its position as an industry leader.

Scania's role in the global truck market continued to evolve as it expanded its presence in emerging markets, particularly in Asia and South America. The company's commitment to providing flexible, customizable solutions for customers in different regions allowed it to tailor its products to meet the specific needs of these markets. Whether it was developing trucks for long-haul operations in North America, construction vehicles for emerging economies, or city trucks for urban environments, Scania's ability to adapt to regional demands made it a trusted partner worldwide.

The company also embraced the growing trend of digitalization in the transport industry, offering **connected services** that allowed fleet operators to monitor the performance of their vehicles in real time. This focus on connectivity further enhanced Scania's appeal to customers, as it provided them with valuable insights into vehicle performance, maintenance needs, and fuel efficiency.

In the years to come, Scania's role in the global truck market would continue to evolve, with a focus on sustainability, technological innovation, and customer-centric solutions. As the company moved into the next decade, it was well-positioned to continue its leadership in the transport industry, driving the transition to a more sustainable future for road transport.

Conclusion

The early 21st century marked a period of significant progress and transformation for Scania. The company's focus on sustainability, technological innovation, and fuel efficiency helped it navigate a rapidly changing market and adapt to the growing demands for cleaner, more efficient vehicles. With the introduction of advanced engine technologies, improved safety features, and the development of the Euro 3 and Euro 4 emission-compliant trucks, Scania reinforced its commitment to reducing emissions and enhancing fuel economy. As the company continued to expand its global presence and embrace new technologies, it solidified its position as a leader in the global truck market, ready to meet the challenges and opportunities of the future.

Chapter 10: Scania's Digital Revolution

The rapid pace of digitalization over the past two decades has transformed virtually every industry, and the trucking sector is no exception. For Scania, embracing digital technologies was not just about keeping up with the times but about staying ahead of the curve and enhancing the way vehicles were designed, operated, and managed. The introduction of digital solutions, including connected trucks, fleet management systems, and telematics, revolutionized the way Scania engaged with its customers, drivers, and fleet operators. These innovations played a crucial role in improving operational efficiency, reducing costs, and enhancing safety and driver comfort, allowing Scania to further solidify its position as an industry leader in the modern era.

The Impact of Digitalization on the Trucking Industry

The digital revolution in the trucking industry was driven by the growing demand for greater efficiency, improved safety, and better control over operations. As fleets became larger and more complex, the need for digital tools to streamline operations and enhance performance became more apparent. Digital technologies began to play a key role in helping fleet operators monitor and manage their vehicles, optimize routes, and ensure that trucks were being used to their fullest potential.

Scania embraced this shift early on, understanding that digital solutions were not just an optional addition but an integral part of the future of transportation. By integrating digital technologies into its vehicles, Scania was able to provide real-time data that could be used to make informed decisions about everything from vehicle performance to maintenance schedules. This was a major leap forward in terms of efficiency, as it allowed operators to manage their fleets proactively rather than reactively.

One of the most significant aspects of this digital transformation was the ability to connect trucks to the internet, enabling the transmission of data in real-time. This was the beginning of the development of **connected trucks**, which would form the foundation for a new era of intelligent transportation. The ability to remotely monitor and control vehicles opened up new opportunities for improving fleet performance, optimizing routes, and reducing fuel consumption.

Development of Connected Trucks and Fleet Management Solutions

As digital technologies evolved, Scania began developing **connected trucks** that could communicate with fleet operators, service centers, and the broader logistics network. These trucks were equipped with sensors, GPS tracking, and onboard diagnostics, allowing them to send real-time data to central management systems. By connecting trucks to the cloud, Scania enabled fleet operators to monitor vehicle performance, fuel consumption, and maintenance needs from anywhere in the world.

The development of connected trucks also led to the creation of **fleet management solutions** that allowed businesses to optimize their operations. Scania's fleet management system offered a range of tools designed to improve fleet efficiency, from route planning and scheduling to performance monitoring and predictive maintenance. These systems provided fleet operators with comprehensive data about their vehicles, including location, fuel usage, tire pressure, engine health, and even driving behavior. This wealth of information helped operators make smarter decisions, streamline logistics, and reduce downtime.

By collecting data from thousands of trucks across the globe, Scania was able to refine its systems continuously, ensuring that its fleet management solutions became increasingly sophisticated over time. Fleet operators could receive real-time alerts about issues such as engine malfunctions or when maintenance was due, allowing for preventive measures to be taken before problems escalated. This predictive approach to maintenance not only extended the life of the vehicles but also helped operators save money by avoiding costly repairs and reducing downtime.

These developments allowed Scania to offer more than just trucks; it was now providing a complete transportation solution, one that integrated hardware, software, and services to help fleet operators improve the efficiency and performance of their operations.

The Role of Telematics and Data Analytics in Improving Efficiency and Reducing Costs

At the heart of Scania's digital revolution was the use of **telematics** and **data analytics**. Telematics refers to the technology used to transmit data between a vehicle and a remote server, and it played a central role in enabling the connected trucks that were at the forefront of Scania's digital solutions. Through telematics, Scania was able to capture a vast array of data from its trucks, which was then analyzed to gain valuable insights into vehicle performance, driver behavior, and fleet operations.

Data analytics turned this data into actionable information. By analyzing patterns in the data, Scania was able to offer fleet operators insights that helped them improve efficiency and reduce costs. For example, by tracking fuel consumption data, Scania could identify opportunities to optimize routes and reduce idle times, which helped operators save fuel and cut operational costs. Telematics also allowed for the optimization of maintenance schedules. By monitoring the health of key components, such as the engine, transmission, and tires, Scania could provide recommendations for maintenance at the optimal time, ensuring that vehicles were running at peak efficiency.

In addition to improving fuel efficiency, data analytics also played a crucial role in reducing wear and tear on vehicles. Scania's fleet management system could track driving behaviors, such as harsh braking, rapid acceleration, and speeding, which not only impacted fuel consumption but also contributed to premature wear of critical components. By providing drivers with feedback on their driving habits, Scania helped them make adjustments to improve their performance, resulting in lower operating costs and fewer maintenance issues.

Furthermore, the insights gained from telematics and data analytics enabled Scania to refine its vehicle designs. The company used real-world data to assess how its trucks were performing under different conditions, allowing it to make design improvements that further enhanced efficiency, durability, and safety.

Scania's Efforts to Enhance Driver Safety and Comfort Through Technology

The digital revolution also had a profound impact on **driver safety and comfort**, an area that Scania had always been committed to improving. As trucks became more advanced and capable, ensuring that drivers could operate them safely and comfortably became a key priority. Scania integrated a wide range of technologies aimed at making the driving experience safer, more comfortable, and more efficient.

Driver assistance systems played a central role in improving safety. Technologies such as **Adaptive Cruise Control (ACC)**, **Lane Departure Warning (LDW)**, and

Automatic Emergency Braking (AEB) became standard features on Scania trucks. These systems worked together to assist drivers in maintaining safe speeds, staying in their lanes, and preventing accidents. The goal was not to replace the driver, but to support them in making better decisions and responding more quickly to potential hazards.

Scania's commitment to driver comfort also extended to the cabin. The company continued to innovate in cab design, ensuring that drivers had a comfortable and ergonomic environment for long journeys. Features such as improved seat designs, better air conditioning, noise reduction, and more intuitive controls helped create a space that reduced fatigue and improved overall well-being. The introduction of **driver-facing cameras** and other monitoring systems allowed Scania to assess driver behavior in real-time, ensuring that drivers remained alert and focused on the road.

Furthermore, Scania introduced the **Scania Driver Support System**, a system that provided real-time feedback to drivers on their performance. It offered coaching tips to improve driving habits, such as optimal shifting, braking, and acceleration techniques. This not only helped reduce fuel consumption but also improved safety by promoting smoother and more controlled driving.

Scania's approach to driver safety and comfort was rooted in its belief that a well-supported driver was essential to maintaining high safety standards and ensuring the longevity of the vehicle. By integrating advanced technologies and continuously improving the driving environment, Scania helped drivers perform at their best while reducing the likelihood of accidents and enhancing their overall experience.

Conclusion

Scania's digital revolution transformed the way the company operated and interacted with its customers, from the development of connected trucks to the use of telematics and data analytics. The integration of these technologies enabled fleet operators to improve efficiency, reduce costs, and enhance safety. At the same time, Scania's efforts to improve driver safety and comfort ensured that the human aspect of trucking was not overlooked. As digitalization continued to shape the future of the transport industry, Scania's commitment to innovation, sustainability, and customer-centric solutions positioned it to lead the way in the next phase of the trucking revolution. With a focus on connected services, predictive maintenance, and driver support, Scania embraced the digital age, ensuring its continued success in an increasingly connected and automated world.

Chapter 11: Scania's Electric Future: Green Transport and Innovation

As the world grapples with the urgent need to combat climate change, the transport industry has come under increasing pressure to reduce its environmental impact. Scania, which has long been a leader in the heavy-duty truck market, recognized early on that sustainability would be key to its future. The development of electric trucks and hybrid vehicles became a central part of Scania's strategy to meet both the demands of a changing market and the increasing regulatory requirements aimed at reducing emissions. The shift toward electrification not only marked a new chapter in the company's history but also set Scania on the path to playing a pivotal role in the transition to a more sustainable transport sector.

The Development of Electric Trucks and Hybrid Vehicles

Scania's journey into electric and hybrid vehicles began with a commitment to reduce the carbon footprint of the transportation industry. With the growing recognition that electric mobility would be crucial to achieving sustainability targets, Scania started investing heavily in research and development of electric powertrains. The goal was to create a range of electric trucks capable of handling the heavy-duty demands of the logistics industry while offering a greener alternative to conventional diesel-powered trucks.

In the early stages, Scania focused on hybrid solutions, combining electric and diesel power to create more fuel-efficient vehicles. The hybrid vehicles could operate in purely electric mode for short distances, especially in urban areas, and switch to the diesel engine for longer trips, where the electric range was not sufficient. This offered a practical solution for fleet operators who were looking to reduce emissions in city centers without compromising on long-haul capabilities.

Scania's hybrid trucks offered numerous benefits, including reduced fuel consumption, lower CO_2 emissions, and improved overall energy efficiency. The company introduced several hybrid models, including the **Scania P-series Hybrid**, which was designed for urban operations. These trucks were particularly well-suited for municipalities and urban logistics companies that needed a combination of environmental performance and operational flexibility. By combining electric propulsion for city driving with a traditional diesel engine for highway journeys, Scania provided a sustainable solution that met the needs of a diverse range of customers.

As Scania gained experience with hybrid technology, it became clear that fully electric trucks would be the next step in the transition to sustainable transport. In the late 2010s, Scania began developing **fully electric heavy-duty trucks**, with the aim of creating vehicles that could replace traditional diesel-powered trucks in urban, regional, and long-distance transport operations. These electric trucks would be powered by high-capacity batteries capable of supporting longer distances while offering zero emissions.

Scania's Role in the Transition to Sustainable Transport

Scania's commitment to sustainability extended beyond the development of electric vehicles. The company understood that the electrification of the transport sector was not just about producing electric trucks but also about contributing to the wider transformation of the industry toward green, low-emission transport solutions. Scania's approach to sustainability focused on **holistic solutions** that included energy-efficient vehicles, renewable energy sources, and the integration of alternative fuels.

Scania has been an active player in pushing for the transition to **sustainable transport systems**, working closely with governments, regulatory bodies, and customers to develop policies that support the adoption of electric and hybrid vehicles. The company was involved in various pilot programs aimed at demonstrating the viability of electric trucks in real-world conditions, from urban deliveries to regional hauls.

In addition to vehicles, Scania also worked on developing the necessary infrastructure to support electric transport. The development of charging networks, for example, became a critical component in the adoption of electric trucks. Scania has partnered with energy providers and other stakeholders to create a reliable and efficient charging infrastructure that would make it easier for fleet operators to incorporate electric trucks into their fleets.

Scania's focus on green transport solutions extended to all aspects of its business, including its production processes. The company set ambitious sustainability targets for its own operations, reducing its carbon emissions, optimizing energy consumption in its factories, and ensuring that the vehicles it produced met the highest environmental standards.

Key Models in the Electric and Hybrid Ranges

As part of its efforts to lead the way in electrification, Scania launched several key models in its electric and hybrid ranges. One of the earliest electric vehicles in Scania's portfolio was the **Scania Citywide Low Entry Hybrid Bus**, introduced in the mid-2010s. Designed for urban environments, the hybrid buses were equipped with a combination of a conventional diesel engine and an electric motor. This hybrid design allowed the buses to operate in electric mode within city centers, reducing air pollution, noise, and congestion. The Citywide buses became a popular choice for cities looking to meet environmental goals while maintaining the performance of public transport systems.

Building on the success of its hybrid buses, Scania introduced its first fully electric truck, the **Scania R 450 Hybrid**, in 2018. The vehicle was part of a pilot project aimed at demonstrating the viability of electric trucks for heavy-duty operations. The R 450 Hybrid could run on electric power for shorter distances and seamlessly transition to its diesel engine for longer distances, providing the benefits of electric propulsion without compromising on the vehicle's range or performance.

In 2020, Scania launched the **Scania L-series electric truck**, which was designed for inner-city deliveries and short-haul transport. This fully electric truck was equipped with batteries capable of providing a range of up to 250 kilometers (155 miles) on a single charge, making it ideal for urban logistics. The Scania L-series was equipped with the latest electric drivetrains, ensuring high energy efficiency and minimal environmental impact. With a focus on urban operations, the L-series offered a quiet, zero-emissions solution for fleets operating in increasingly environmentally regulated city environments.

Scania continued to innovate with the development of electric trucks suited for longer distances. The company's **Scania R-series Electric**, introduced in 2021, offered a greater range and was designed to meet the needs of regional transport. The truck's battery capacity allowed for longer trips without the need for frequent charging stops, making it a viable alternative for long-haul routes in areas where charging infrastructure was developing.

Challenges and Opportunities in Electrifying the Heavy Transport Sector

While the electrification of the transport sector represents a tremendous opportunity for both Scania and the industry as a whole, it also presents several challenges. One of the biggest hurdles facing the widespread adoption of electric trucks is the **high cost of batteries**. The cost of lithium-ion batteries, while steadily decreasing, remains a significant factor in the overall price of electric trucks, making them more expensive than their diesel counterparts. For many fleet operators, this price difference is a barrier to adoption, even though the long-term operational savings from lower fuel and maintenance costs can offset the initial investment.

The **charging infrastructure** required to support the large-scale adoption of electric trucks is another challenge. While urban areas and some highways are beginning to see the development of charging networks, the infrastructure needed to support electric heavy-duty trucks on a broader scale is still lacking in many regions. This challenge is particularly acute in remote or rural areas, where long-distance travel is required.

However, Scania sees these challenges as opportunities to innovate. By partnering with infrastructure providers, Scania is helping to drive the development of charging networks and working to bring down the cost of batteries through research and development. The company is also exploring **alternative energy sources**, such as hydrogen fuel cells, which could offer an additional route to sustainable transport solutions, particularly for long-haul operations where battery limitations are more pronounced.

The growing interest in **sustainable transport** also presents Scania with new market opportunities. Governments around the world are setting stricter emissions standards and providing incentives for companies that adopt green technologies. Scania's electric and hybrid vehicles position the company well to capitalize on these trends, particularly in regions like Europe, where there is a strong push toward zero-emissions transport solutions.

Conclusion

Scania's transition to electric and hybrid vehicles represents a critical step in the company's long-term strategy to become a leader in sustainable transport. By investing in electric truck development, creating a range of hybrid solutions, and working to address the challenges of electrification, Scania is positioning itself at the forefront of the green transport revolution. The company's commitment to sustainability is reflected not just in the vehicles it produces but in its efforts to transform the entire transport ecosystem, from infrastructure to regulatory support. As Scania continues to innovate and lead in the electric and hybrid vehicle market, it will play a key role in the transition to a cleaner, more sustainable future for global transport.

Chapter 12: The Role of Scania in the Global Supply Chain

The trucking industry is an essential component of the global supply chain, with transportation networks facilitating the movement of goods across countries and continents. As one of the world's leading manufacturers of heavy-duty trucks, Scania has played a pivotal role in shaping the efficiency, reliability, and sustainability of these supply chains. Over the years, the company has not only built a strong reputation for producing powerful and durable vehicles but also developed strategic partnerships with logistics companies and fleet operators to enhance the flow of goods globally. Scania's commitment to innovation, fuel efficiency, and sustainability has enabled it to contribute significantly to the optimization of the world's transportation networks.

Scania's Partnerships with Logistics Companies and Fleet Operators

From the early days of its expansion into international markets, Scania has recognized the importance of collaboration with logistics companies and fleet operators. These partnerships have been key to understanding the specific needs of the global supply chain and designing trucks that meet the demands of industries ranging from long-haul freight to urban logistics. Scania's approach to partnership has always been one of shared goals: improving efficiency, reducing costs, and enhancing the sustainability of transportation networks.

Scania has worked closely with large fleet operators, including logistics giants such as **DHL**, **Maersk**, and **XPO Logistics**, to provide tailored solutions that address their unique challenges. By engaging directly with these companies, Scania has been able to better understand the operational realities of running a fleet at a global scale. This has led to the development of trucks optimized for specific functions, such as urban deliveries, cross-border shipping, or long-haul transportation.

For example, Scania's **connected fleet management solutions** have played a crucial role in helping logistics companies optimize their operations. By using telematics and real-time data, Scania's fleet management systems allow logistics companies to track the location and performance of their trucks, monitor fuel consumption, and schedule maintenance more efficiently. This has helped fleet operators reduce downtime, improve fuel efficiency, and optimize routes, all of which contribute to a more efficient supply chain.

Additionally, Scania's collaboration with fleet operators has provided valuable insights into the challenges of modern logistics. As e-commerce continues to grow and demand for faster deliveries increases, logistics companies are turning to Scania for vehicles that can operate efficiently in an urban environment while still maintaining the ability to handle long-haul distances. By adapting its vehicle designs to meet these diverse needs, Scania has established itself as a key partner for fleet operators looking to stay competitive in a rapidly changing industry.

Contributions to Global Supply Chain Efficiency Through Innovation

Scania has long been at the forefront of developing innovative technologies aimed at improving supply chain efficiency. The company's commitment to research and development has led to numerous advancements in vehicle performance, fuel efficiency, and sustainability, all of which have had a direct impact on the global supply chain.

One of Scania's most significant contributions has been the development of **fuel-efficient engines**. As fuel costs continue to represent a significant portion of operating expenses for logistics companies, Scania's efforts to reduce fuel consumption have directly improved the bottom line for fleet operators. By introducing advanced engine technologies, such as turbocharging and intercooling, Scania has been able to deliver more power with less fuel. This not only reduces operational costs but also helps companies reduce their carbon footprint, which is increasingly important in a world where sustainability is a growing concern.

In addition to engine efficiency, Scania's efforts to improve **aerodynamics** and reduce vehicle weight have also contributed to supply chain efficiency. The company's trucks are designed with features such as **streamlined cabs** and **low-drag chassis**, which reduce fuel consumption by decreasing wind resistance. Lighter trucks also allow for higher payloads, improving the efficiency of each trip and reducing the number of vehicles needed to transport goods.

Scania has also been a leader in **digitalization** within the supply chain. Its fleet management solutions, which integrate telematics, GPS, and data analytics, allow fleet operators to optimize their vehicle routes, monitor driving behavior, and predict maintenance needs. These digital solutions have been a game changer for the logistics industry, providing operators with the tools they need to make data-driven decisions and run their operations more efficiently. The ability to track the condition of vehicles in real-time also helps reduce maintenance costs, minimize downtime, and extend the lifespan of trucks.

The Company's Impact on Infrastructure and Transportation Networks Worldwide

While Scania has always been focused on delivering the best possible vehicles, the company has also recognized the importance of infrastructure in ensuring the smooth flow of goods across global supply chains. As the demand for freight transportation continues to grow, the need for robust and efficient infrastructure has never been greater. Scania has played a key role in shaping the development of transportation networks worldwide, particularly in areas where new infrastructure is being built to meet the demands of a rapidly globalizing economy.

In many regions, Scania has worked in partnership with governments, local authorities, and infrastructure developers to improve transportation networks. For example, Scania has been involved in several initiatives aimed at improving the **charging infrastructure for electric trucks**, enabling the transition to more sustainable transport solutions. As cities and countries set ambitious targets for reducing emissions, the need for reliable charging infrastructure has become increasingly important. Scania's commitment to building the infrastructure necessary to support electric vehicles has made it a key player in the push toward greener transportation systems.

Furthermore, Scania has been involved in initiatives aimed at improving **intermodal transportation**, which combines different modes of transport—such as road, rail, and sea—to create more efficient and sustainable logistics networks. By developing trucks that are compatible with intermodal systems, Scania has made it easier for goods to be transported seamlessly across various modes of transport, reducing transit times, fuel consumption, and overall environmental impact.

Scania has also worked to improve **urban transportation networks** by introducing vehicles that are specifically designed for city deliveries. As cities become more congested and face increasing pressure to reduce pollution, Scania's electric and hybrid trucks have provided a sustainable solution for urban logistics. The development of smaller, more maneuverable trucks that can handle tight city streets has helped improve the efficiency of last-mile delivery operations, which are often the most complex and costly part of the supply chain.

Scania's Role in Providing Transport Solutions for a Rapidly Globalizing Economy

As the global economy has become more interconnected, the demand for efficient, reliable, and cost-effective transport solutions has increased. Scania's ability to adapt to the changing needs of the global supply chain has made it an essential partner for businesses and governments alike. The company's commitment to innovation, sustainability, and efficiency has enabled it to provide solutions that meet the needs of a rapidly globalizing world.

One of the key challenges of the 21st century is the need to transport goods across vast distances while minimizing environmental impact. Scania has been at the forefront of developing solutions that balance these competing demands. By focusing on **sustainability**, Scania has introduced a range of vehicles that use **alternative fuels** and **electric powertrains** to reduce emissions while still meeting the performance needs of the logistics industry. The transition to electric trucks, combined with Scania's development of biofuel-powered vehicles, ensures that the company can continue to meet the demands of a low-carbon global economy.

Moreover, Scania's role in **global supply chains** has expanded beyond just providing vehicles. Through its **connected services**, fleet management systems, and involvement in infrastructure projects, Scania has positioned itself as a key enabler of the efficient movement of goods worldwide. As the logistics sector continues to evolve, Scania's commitment to developing smarter, more efficient, and more sustainable transport solutions will ensure that it remains a driving force in the global supply chain for years to come.

Conclusion

Scania's contributions to the global supply chain go beyond just the production of vehicles; the company has been an active participant in shaping the logistics industry through its innovations, partnerships, and commitment to sustainability. By working closely with fleet operators, logistics companies, and infrastructure developers, Scania has played a key role in improving supply chain efficiency and supporting the transition to more sustainable transportation networks. As the global economy continues to evolve, Scania's ability to adapt and provide solutions that meet the challenges of a rapidly changing world ensures its place at the forefront of the transport industry, ready to lead the way in the next era of global logistics.

Chapter 13: The Driver's Experience: Safety, Comfort, and Technology

The role of the truck driver has always been critical to the success of the transport industry. As the demands on drivers increased and the industry evolved, so too did the need for vehicles that not only offered top performance but also provided a comfortable, safe, and supportive environment for the people behind the wheel. Scania, as one of the world's leading manufacturers of heavy-duty trucks, has long understood that the driver's experience is an essential component of the overall performance and longevity of a vehicle. The company has continuously innovated in the areas of driver comfort, ergonomics, and safety, ensuring that its trucks are not only technologically advanced but also designed with the well-being of the driver in mind.

Scania's Focus on Driver Comfort, Ergonomics, and Safety Features

In the past, truck cabins were often seen as functional but somewhat austere environments designed simply to house the driver. However, Scania recognized that drivers spend long hours on the road, often facing difficult working conditions, and that the cabin should be a place of comfort, efficiency, and safety. As part of its commitment to improving the driver experience, Scania focused on three core areas: **comfort**, **ergonomics**, and **safety**.

Comfort in the truck cabin was a key consideration in Scania's design philosophy. Over the years, the company introduced a range of features designed to make the cabin a more pleasant and functional workspace. High-quality seating, for example, became a priority. Scania's **driver seats** are adjustable in multiple ways, allowing drivers to find the optimal position for comfort and support. The seats are equipped with advanced suspension systems to reduce vibration and minimize fatigue, ensuring that drivers can remain focused during long shifts.

The ergonomic design of the cabin is another area where Scania has excelled. The controls and dashboard are designed for ease of use, with all key functions placed within easy reach. This design minimizes driver distraction and maximizes efficiency, allowing drivers to operate the truck with ease and comfort. Scania's **multi-functional steering wheels**, for example, provide controls for several key features, such as cruise control, multimedia, and phone connectivity, allowing drivers to stay focused on the road while still having access to the necessary tools.

Safety features have always been central to Scania's approach to truck design. The company's commitment to making life on the road safer for drivers has led to the introduction of a range of advanced safety technologies that help protect the driver in the event of an accident and prevent accidents from happening in the first place. Scania's **crash-resistant cabins** are designed to protect the driver in case of a collision, with reinforced structures and strategically placed airbags. These cabins are rigorously tested to meet the highest safety standards, ensuring that drivers are as safe as possible.

Innovations in Cab Design and Interior Technology

Scania's cab design has evolved significantly over the years, reflecting the changing needs of the industry and advancements in technology. The company has consistently pushed the boundaries of what is possible in terms of comfort, functionality, and aesthetics, resulting in a series of cabins that are as much about quality of life as they are about performance.

The introduction of the **Scania R-series cab** marked a major step forward in the evolution of truck interiors. The R-series cabin was designed with a focus on comfort and functionality, incorporating high-end materials and finishes to create a premium space for drivers. The spacious interior features a fully adjustable driver's seat, a wide range of storage compartments, and a modern dashboard with an intuitive layout. This made the R-series cab one of the most comfortable and functional truck interiors available at the time.

Scania's innovation in interior technology also played a crucial role in enhancing the driver experience. The company was one of the first in the industry to integrate advanced **infotainment systems** into its trucks. The **Scania Driver Support System** is a prime example of this innovation. This system provides real-time feedback to drivers on their driving habits, offering suggestions to improve fuel efficiency, reduce wear and tear on the truck, and enhance overall safety. The system also includes features such as **navigation, voice-activated controls,** and **hands-free connectivity**, allowing drivers to stay connected while keeping their focus on the road.

The **sleeping area** in Scania's cabins was also designed with driver comfort in mind. Long-haul drivers often spend nights in their trucks, so Scania equipped its vehicles with a comfortable bed and ample storage space for personal belongings. The layout of the sleeping area is designed to maximize space and ensure that drivers can rest comfortably during their breaks. Scania's attention to detail in the design of these areas reflects the company's recognition that driver well-being extends beyond just the hours spent behind the wheel.

Advanced Driver Assistance Systems (ADAS)

As part of its ongoing commitment to driver safety, Scania has integrated a range of **Advanced Driver Assistance Systems (ADAS)** into its vehicles. These systems are designed to assist drivers in a variety of ways, from maintaining safe distances to detecting potential hazards on the road. ADAS technologies work together to reduce the risk of accidents and help drivers operate their vehicles more safely and efficiently.

One of the key components of Scania's ADAS is **adaptive cruise control (ACC)**, which automatically adjusts the truck's speed to maintain a safe distance from the vehicle ahead. This system uses radar and sensors to monitor traffic conditions, allowing the truck to slow down or speed up as necessary. ACC reduces the risk of rear-end collisions, especially in heavy traffic, and helps the driver maintain a consistent speed over long distances, improving fuel efficiency.

Another important feature of Scania's ADAS is the **lane departure warning system**, which alerts the driver if the truck begins to drift out of its lane without signaling. This system uses cameras to monitor the road markings and can provide visual or audible warnings when the truck veers off course. This feature is particularly useful on long, straight roads, where driver fatigue can cause lapses in concentration.

Collision avoidance systems are also an essential part of Scania's ADAS suite. These systems use radar and cameras to detect objects or other vehicles in the truck's path and can automatically apply the brakes to avoid or mitigate a collision. This feature provides an additional layer of protection in situations where the driver may not have enough time to react to an imminent hazard.

In addition to these safety technologies, Scania's ADAS also includes features such as **automatic emergency braking (AEB)**, **blind-spot detection**, and **driver alertness monitoring**, all of which contribute to safer driving and fewer accidents on the road. By integrating these advanced technologies, Scania ensures that drivers have the support they need to operate their vehicles more safely and efficiently.

Scania's Approach to Making Life on the Road Safer and More Comfortable

At the heart of Scania's approach to improving the driver experience is the recognition that a comfortable, safe, and well-equipped driver is essential to the success of any transport operation. The company's focus on **driver well-being** extends beyond just the vehicle itself and includes support for drivers on and off the road.

One of the key ways Scania makes life on the road safer and more comfortable is through its **driver training programs**. Scania offers a range of training courses designed to help drivers improve their driving skills, increase fuel efficiency, and reduce accidents. The company also offers **driver health and wellness initiatives**, which promote healthy lifestyles for truck drivers, who often face long working hours and challenging conditions. These programs are aimed at improving both physical and mental well-being, ensuring that drivers are fit, healthy, and focused while on the road.

Scania's focus on comfort and safety also extends to the **truck's interior lighting**, which is designed to create a relaxing and functional environment for drivers. Adjustable lighting allows drivers to choose the level of brightness that suits their needs, whether for working, relaxing, or resting. The use of natural materials and finishes in the cabin also contributes to a pleasant and calming atmosphere, reducing stress and promoting relaxation during long shifts.

In addition to the physical aspects of the truck cabin, Scania also ensures that drivers are supported through **digital services**. The **Scania Driver Support System** offers real-time feedback on driving behavior, while the company's fleet management solutions provide operators with data on driver performance, fuel consumption, and safety metrics. This data can be used to coach drivers, helping them improve their performance, reduce fuel consumption, and make their driving habits more sustainable.

Conclusion

Scania's commitment to enhancing the driver experience is a key part of the company's overall strategy for success. By focusing on comfort, safety, and technology, Scania has created a range of trucks that not only perform at the highest level but also provide drivers with a supportive, comfortable, and safe environment. The integration of advanced driver assistance systems, ergonomic design, and cutting-edge technologies has set Scania apart as a leader in the heavy-duty truck market. As the industry continues to evolve, Scania's focus on the driver will remain a critical factor in ensuring the success and sustainability of the global transport sector.

Chapter 14: Scania's Legacy in Motorsports and Performance

Scania's commitment to performance and innovation extends beyond the realm of commercial transportation and into the world of motorsports. Over the years, the company has made significant contributions to various racing disciplines, including rally and truck racing, where its engineering prowess and dedication to high-performance vehicles have been showcased in some of the most challenging competitions on the planet. Scania's involvement in motorsports has not only demonstrated the company's technical capabilities but also influenced the design of its commercial products, bringing lessons learned on the racetrack to the trucks and vehicles that power industries around the world.

Scania's Involvement in Motorsports, Including Rally and Truck Racing

Scania's motorsport legacy began in the early years of the company, with a strong focus on developing vehicles that could perform under extreme conditions. The company's trucks and engines, known for their robustness and reliability, were seen as ideal candidates for various motorsport events, where performance and durability are paramount.

One of Scania's most notable motorsport endeavors was its participation in **truck racing**, particularly in the **European Truck Racing Championship (ETRC)**. This high-octane form of motorsport pits powerful trucks against one another in competitive races that take place on circuits designed to challenge both the driver and the vehicle. Scania's involvement in truck racing allowed the company to showcase its trucks' performance capabilities, especially their strength, handling, and speed. The truck racing events also provided a testing ground for Scania's innovations in braking, suspension, and engine management, as these features were constantly pushed to their limits in the heat of competition.

Scania's trucks have had considerable success in the European Truck Racing Championship, often competing against other major manufacturers such as Mercedes-Benz, MAN, and Volvo. The competition has allowed Scania to demonstrate how its trucks, built for heavy-duty performance, could also excel on the racetrack. By participating in truck racing, Scania reinforced its brand's identity as a manufacturer of both powerful and reliable vehicles.

In addition to truck racing, Scania has been involved in **rally racing**, where its vehicles have competed in some of the most grueling off-road competitions in the world. Scania's rugged and durable trucks have been well-suited for rally events, where vehicles are subjected to harsh terrain, extreme weather, and challenging obstacles. These events have provided Scania with valuable insights into vehicle durability, handling, and performance under demanding conditions, lessons that have been applied to the company's commercial products.

Performance-Oriented Vehicle Designs for Competition

Scania's success in motorsports can be attributed to its relentless pursuit of performance-oriented vehicle designs. Motorsports, with its emphasis on speed, agility, and durability, has provided a platform for Scania to test and refine its engineering solutions, resulting in trucks that are not only high-performing but also capable of withstanding the pressures of commercial use.

One of the key aspects of Scania's motorsport-inspired designs is **engine performance**. The company has continuously developed and refined its engines for motorsports, pushing the limits of power and efficiency.

Scania's participation in truck racing has driven the development of more advanced engine technologies, including **turbocharged systems**, **fuel injection management**, and **aerodynamic improvements**, all of which have helped improve vehicle performance on the racetrack and in real-world conditions. The lessons learned in motorsports, particularly in the areas of engine power, cooling systems, and fuel efficiency, have directly influenced the performance of Scania's commercial trucks, making them some of the most powerful and efficient vehicles on the road today.

The development of **suspension systems** is another area where motorsports have played a crucial role in Scania's vehicle designs. In truck racing, the suspension system is critical for managing the stresses placed on the vehicle during high-speed corners, jumps, and rough surfaces. Scania's experience in rally and truck racing has allowed the company to develop suspension systems that provide superior stability, handling, and comfort, even under the most challenging conditions. These advancements have found their way into Scania's commercial products, contributing to a smoother driving experience and improved cargo handling in real-world applications.

In addition to performance under extreme conditions, **aerodynamics** has played a significant role in Scania's motorsport vehicles. On the racetrack, the ability to reduce drag while maximizing stability is essential for achieving high speeds and maintaining control. Scania's expertise in aerodynamic design, honed in motorsports, has influenced the development of more fuel-efficient and stable commercial trucks, ensuring that Scania's vehicles are as efficient on the road as they are on the racetrack.

The Influence of Motorsports on Scania's Commercial Products

The lessons learned through Scania's motorsport involvement have had a direct and lasting impact on the company's commercial products. The competitive pressure of motorsport forces constant innovation and improvement, and Scania has been able to take the performance-enhancing technologies and strategies developed in racing and apply them to its line of commercial trucks and vehicles.

For instance, **engine technology** developed for Scania's racing vehicles has been adapted for use in the company's commercial trucks. The turbocharged engines, high-performance fuel management systems, and advanced cooling mechanisms tested on the racetrack have all been incorporated into Scania's production vehicles. These innovations help Scania's trucks achieve higher fuel efficiency, better power-to-weight ratios, and improved performance, which ultimately benefits fleet operators by reducing operating costs and improving vehicle longevity.

The development of **lightweight materials** used in Scania's motorsport vehicles has also influenced the design of the company's commercial trucks. By using advanced materials to reduce vehicle weight while maintaining strength and durability, Scania has been able to improve fuel efficiency and payload capacity, both of which are essential for long-haul trucking operations.

Moreover, Scania's expertise in **braking technology** developed for motorsports has translated into the company's **heavy-duty braking systems** for commercial vehicles. The precision and power of Scania's racing brakes, designed to withstand extreme temperatures and pressures, have influenced the development of high-performance braking systems in Scania trucks, ensuring they can handle heavy loads and challenging road conditions safely and efficiently.

Key Milestones in Scania's Motorsport History

Scania's motorsport history is marked by several key milestones that have highlighted the company's engineering capabilities and performance-focused mindset.

One of the standout moments in Scania's motorsport legacy was its **first major success in truck racing** in the 1980s. Scania entered the European Truck Racing Championship, competing against other heavy-duty truck manufacturers. This marked the beginning of a long and successful involvement in the sport, where Scania's trucks demonstrated their power, durability, and handling, setting the stage for the company's ongoing participation in the championship.

In the 1990s, Scania achieved significant success in truck racing, with its **R-series trucks** becoming iconic in the European Truck Racing Championship. The success of Scania's trucks in these races not only showcased the company's engineering expertise but also helped solidify its reputation as a leader in performance and innovation.

Another key milestone was Scania's **involvement in rally racing**, where the company's trucks competed in some of the most challenging off-road races, including the **Paris-Dakar Rally**. Scania's vehicles excelled in these harsh environments, where durability, strength, and off-road capability were critical to success. The lessons learned from these rally races contributed to Scania's understanding of vehicle durability and performance under extreme conditions, helping shape the design of its commercial products.

Conclusion

Scania's legacy in motorsports is a testament to the company's commitment to performance, innovation, and engineering excellence. The company's participation in truck racing, rally events, and other motorsport disciplines has not only showcased its vehicles' capabilities but also provided valuable insights that have influenced the design of Scania's commercial trucks. The technologies developed and refined on the racetrack—ranging from engine performance and suspension systems to aerodynamics and braking—have been integrated into Scania's production vehicles, ensuring that its trucks remain at the cutting edge of performance, reliability, and efficiency. As Scania continues to push the boundaries of what is possible in both motorsports and the commercial vehicle market, its legacy in performance will remain a defining feature of the brand for years to come.

Chapter 15: Scania Today: A Global Leader in Heavy Transport

Scania's journey over the past century has been one of innovation, adaptation, and growth. Today, Scania stands as one of the leading manufacturers of heavy-duty vehicles in the world, renowned not only for its performance and reliability but also for its commitment to sustainability and innovation. As the global transport industry evolves, Scania remains at the forefront, shaping the future of transport with its cutting-edge technologies, diverse product offerings, and focus on providing sustainable, efficient solutions to meet the demands of a rapidly changing world.

Overview of Scania's Current Position in the Global Market

Scania is one of the largest manufacturers of heavy-duty trucks, buses, and engines in the world, operating in over 100 countries and employing thousands of people across its global network. The company's reputation for quality and innovation has earned it a loyal customer base, from fleet operators and logistics companies to municipal transport systems and industries requiring specialized vehicles.

Scania's strong global position is supported by its diverse product offerings, extensive service network, and commitment to continuous improvement. The company has built a reputation for providing high-performance vehicles that offer low total cost of ownership, exceptional durability, and minimal environmental impact. As of today, Scania has a broad product range that spans heavy-duty trucks, buses, and industrial engines, with a strong focus on high-performance solutions tailored to specific market needs.

Scania has remained a leader in Europe, where it continues to hold a significant market share, but its expansion into international markets, particularly in Asia, South America, and Africa, has strengthened its global presence. The company has embraced new technologies and alternative energy solutions to remain competitive in these growing markets, where sustainability and efficiency are becoming increasingly important.

Key Product Lines and New Model Introductions

Scania's product portfolio includes a broad range of vehicles designed to meet the diverse needs of industries around the world. Its trucks are known for their durability, fuel efficiency, and innovation, with the company offering solutions for long-haul, regional, and urban transport. Scania's **R-series** and **S-series** heavy-duty trucks, which have set the benchmark for performance, comfort, and safety, continue to be among the company's flagship products. These models offer a variety of configurations, from standard long-haul trucks to specialized vehicles for construction and mining operations.

The **Scania XT-series**, introduced in recent years, is designed specifically for construction and heavy-duty operations in rough, off-road environments. These vehicles are built to withstand the harshest conditions, offering enhanced durability, power, and performance for industries such as construction, quarrying, and forestry.

In addition to its well-established product lines, Scania has made significant strides in electrification and hybrid technologies. The company introduced its first fully electric truck, the **Scania L-series Electric**, designed for urban transport and inner-city deliveries. The L-series Electric combines zero-emission driving with the robust capabilities that Scania trucks are known for. With a focus on sustainable urban transport, this model is a key part of Scania's strategy to provide environmentally friendly solutions for short-haul, city-based logistics.

Scania's commitment to sustainability is also evident in its **hybrid trucks**, which offer the flexibility of combining a conventional diesel engine with an electric motor. These vehicles allow fleet operators to reduce fuel consumption and emissions, particularly in urban environments, without sacrificing the power and range needed for longer trips.

Scania's Focus on Digital Solutions, Connectivity, and Autonomous Vehicles

In addition to its continued innovation in vehicle design and technology, Scania has placed a strong emphasis on **digital solutions**, **connectivity**, and the development of **autonomous vehicles**. These areas of focus are not just about enhancing the performance of Scania's products but also about providing fleet operators and customers with smarter, more efficient ways to manage their operations.

At the core of Scania's digital strategy is its **connected services**, which leverage telematics, real-time data, and analytics to improve fleet performance and reduce costs. Scania's fleet management system, **Scania Fleet**, provides fleet operators with insights into their vehicle operations, including real-time location tracking, fuel consumption data, and maintenance needs. This allows operators to make data-driven decisions to optimize their fleets, improve fuel efficiency, and reduce downtime.

Scania's connected solutions also enable remote diagnostics, allowing fleet operators to identify potential issues before they cause breakdowns, schedule maintenance more effectively, and reduce the risk of costly repairs. The company's **Scania Driver Support System** provides drivers with feedback on their driving habits, helping to improve fuel efficiency, reduce wear and tear on vehicles, and promote safer driving practices.

As part of its digitalization strategy, Scania is also heavily involved in the development of **autonomous vehicles**. The company has been exploring autonomous driving technologies for several years, focusing on developing systems that can improve safety, efficiency, and driver productivity. Scania's autonomous trucks are being designed for long-haul operations, where automation could help optimize routes, reduce fuel consumption, and improve overall efficiency by enabling trucks to operate 24/7.

The development of autonomous vehicles is still in its early stages, but Scania's involvement in the field positions it to be a leader in the next generation of transport technology.

The company's work in this area, combined with its existing expertise in connectivity and digital solutions, ensures that Scania will remain at the forefront of technological innovation in the transport sector.

The Company's Role in Providing Transport Solutions for a Rapidly Globalizing Economy

As the global economy continues to become more interconnected, the demand for efficient, reliable, and sustainable transport solutions has never been greater. Scania, with its focus on performance, innovation, and sustainability, is well-positioned to meet the challenges of this rapidly evolving market.

Scania's trucks and transport solutions play a critical role in **global supply chains**, ensuring that goods move efficiently across borders, industries, and supply networks. The company's vehicles are not only optimized for performance and reliability but are also designed to meet the growing need for sustainability. Scania's investments in **electric**, **hybrid**, and **biofuel-powered trucks** are helping the transport sector transition to a low-carbon future, addressing the global push for reduced emissions and improved air quality.

Scania's ability to provide **customized transport solutions** is one of the reasons it has maintained such a strong position in the global market. The company's product range, which includes trucks for long-haul, urban, and specialized transport, allows Scania to serve diverse industries such as logistics, construction, mining, and municipal transport. By offering tailored solutions that meet the unique needs of these industries, Scania has built strong relationships with fleet operators, logistics companies, and municipalities, ensuring its place as a key player in the global transport sector.

Furthermore, Scania's emphasis on **digitization and connectivity** aligns perfectly with the needs of a global economy that demands greater efficiency, data transparency, and automation. As supply chains become more complex and customer expectations for speed and reliability increase, Scania's digital tools, fleet management systems, and autonomous vehicle development will play an essential role in enabling more efficient, sustainable, and cost-effective transport networks.

Conclusion

Today, Scania is more than just a manufacturer of trucks—it is a leading provider of comprehensive transport solutions that help drive the global economy. With its focus on innovation, sustainability, and performance, Scania has built a solid foundation for continued success in an ever-evolving market. From its advanced product lines and commitment to electrification to its digital solutions and work in autonomous vehicles, Scania is positioning itself to lead the way in the transport sector for years to come. As the company continues to innovate and adapt to the demands of a rapidly globalizing world, Scania's legacy as a global leader in heavy transport is assured.

Chapter 16: The Road Ahead: Scania's Vision for 2030 and Beyond

As the world of transport continues to evolve, Scania has long understood that success in the future depends on its ability to adapt, innovate, and lead the way toward a more sustainable, efficient, and technologically advanced industry. With the global shift toward green transport, digitization, and automation, Scania is positioning itself at the forefront of this transformation, focusing on delivering innovative solutions to meet the challenges of a rapidly changing world. As we look to the future, Scania's strategic goals and vision for 2030 and beyond will define not only the company's path forward but also its role in shaping the future of the global transport sector.

Scania's Long-Term Strategic Goals and Vision for the Future

Scania's vision for 2030 is built on a foundation of three core pillars: **sustainability**, **digitalization**, and **autonomous driving**. The company's overarching goal is to lead the transition to a sustainable, low-carbon transport system, while also continuing to drive technological innovation and improve operational efficiency for its customers. Scania aims to be the preferred partner for its customers, providing not just trucks but comprehensive transport solutions that support their goals in an increasingly complex and competitive market.

To achieve this, Scania has outlined a series of long-term strategic goals focused on enhancing vehicle performance, reducing environmental impact, and leveraging cutting-edge technology to create a more efficient transport ecosystem. By 2030, Scania aims to have a significant portion of its product range electrified or using alternative fuels, reducing emissions across the board and further solidifying its leadership in green transport solutions. The company's efforts to integrate digital solutions, connectivity, and automation into its vehicles will continue to be a central part of its strategy, enhancing fleet management and making transport more efficient, safe, and cost-effective.

Innovations Expected in the Next Decade, Including Autonomous Trucks and Further Electrification

The next decade will see Scania continue to push the boundaries of what is possible in the transport industry. Scania has set ambitious goals for **autonomous vehicles**, envisioning a future where trucks can operate independently over long distances, with minimal human intervention. The development of **autonomous trucks** will be driven by advancements in **machine learning**, **artificial intelligence (AI)**, and **vehicle-to-vehicle communication**, allowing trucks to navigate roads, optimize routes, and communicate with other vehicles and infrastructure in real-time. This technology will significantly reduce human error, improve safety, and increase operational efficiency, allowing trucks to operate around the clock, increasing capacity and reducing delivery times.

Scania's work in autonomous technology is already underway, with several pilot projects focusing on the integration of autonomous trucks in controlled environments, such as closed industrial sites and controlled roadways. However, the broader vision is for **level 4 and 5 autonomy** on public roads, where trucks will be capable of driving themselves in most situations. Scania's goal is to lead the development of autonomous transport solutions for the heavy-duty sector, particularly for long-haul trucking, where autonomous technology can help optimize routes and fuel consumption while ensuring driver safety.

Alongside autonomous driving, **electrification** will continue to be a major focus for Scania in the next decade.

The company has already made significant strides in developing electric trucks, but by 2030, Scania plans to have an expanded range of fully electric vehicles capable of meeting the diverse needs of its customers. This includes the development of electric trucks for long-haul transportation, which is currently one of the most significant challenges for the industry, given the large distances and high power requirements involved. Scania is investing in **high-capacity batteries**, faster charging technology, and more efficient electric drivetrains to make electric long-haul trucks a viable alternative to traditional diesel-powered vehicles.

Scania's hybrid and electric models will be further refined and expanded to cover more applications, from **urban delivery** to **regional haulage**, all while ensuring that the total cost of ownership remains competitive. With the rapid expansion of charging infrastructure, Scania's electric vehicles will become an integral part of the future transport ecosystem, providing fleet operators with zero-emission solutions that reduce their carbon footprint and operating costs.

The Company's Commitment to Sustainability and Its Roadmap to Achieve Carbon Neutrality

Sustainability has long been at the heart of Scania's mission, and as the world faces the growing urgency of addressing climate change, the company has made clear its ambition to achieve **carbon neutrality** by 2050. However, Scania's roadmap toward this goal begins with more immediate targets for 2030. Scania aims to reduce its **carbon footprint** by 20% per vehicle and by 50% across its entire operations by 2030. This will be achieved through a combination of electrification, the use of **alternative fuels**, and improving the energy efficiency of its vehicles and production processes.

Scania's commitment to sustainability also extends to the production side. The company plans to reduce emissions from its factories by transitioning to renewable energy sources, optimizing production processes, and minimizing waste. The use of **sustainable materials** and **circular economy principles** will be central to this effort, ensuring that Scania's vehicles are not only more energy-efficient but also have a lower environmental impact throughout their lifecycle.

Scania's work with alternative fuels, such as **biogas**, **ethanol**, and **hydrogen**, will play a critical role in achieving carbon neutrality. These renewable fuels can significantly reduce emissions compared to traditional diesel and offer a flexible, scalable solution for fleet operators looking to decarbonize their operations. Scania has already developed **biofuel-powered trucks** and is working closely with fuel producers and regulators to ensure that sustainable fuel options become more widely available.

The Future of Transport and Scania's Place Within It

The future of transport is undoubtedly evolving, driven by the need for more sustainable, efficient, and automated solutions. As the world increasingly demands cleaner transport options and seeks to reduce the environmental impact of the logistics industry, Scania is positioning itself as a key enabler of this transformation.

In the future, Scania's role will extend beyond providing vehicles; it will be a comprehensive solutions provider, integrating vehicles, digital services, and energy systems into a seamless transport ecosystem. Through **connected vehicles**, **data analytics**, and **fleet management**

solutions, Scania will help operators optimize their fleets, reduce costs, improve safety, and enhance productivity.

The company's focus on **sustainable urban mobility** will also contribute to the development of cleaner cities. Scania's electric buses, for example, are already making a significant impact on urban public transport systems, reducing emissions and noise pollution while providing a more sustainable alternative to conventional diesel buses. As cities continue to grow and urban congestion becomes more of a concern, Scania's solutions will be central to building smarter, greener cities.

As automation, electrification, and sustainability drive the transport sector forward, Scania's continued commitment to **innovation and customer-centric solutions** will ensure that the company remains at the heart of the global transport network. Whether it's developing fully autonomous long-haul trucks, providing zero-emission solutions for urban logistics, or optimizing supply chain performance through digital tools, Scania's vision for the future is one of leadership, innovation, and sustainability.

Conclusion

Scania's vision for 2030 and beyond is a bold one, focused on achieving carbon neutrality, embracing new technologies, and providing sustainable transport solutions for a rapidly changing world. As the company looks to the future, it is committed to transforming the transport industry with innovations in electrification, autonomy, and digitalization. By continuing to push the boundaries of what is possible in vehicle technology, Scania will ensure its place as a global leader in heavy transport, driving the future of sustainable transport solutions while maintaining its legacy of performance, innovation, and customer satisfaction.

About the Author

Etienne Psaila, an accomplished author with over two decades of experience, has mastered the art of weaving words across various genres. His journey in the literary world has been marked by a diverse array of publications, demonstrating not only his versatility but also his deep understanding of different thematic landscapes. However, it's in the realm of automotive literature that Etienne truly combines his passions, seamlessly blending his enthusiasm for cars with his innate storytelling abilities.

Specializing in automotive and motorcycle books, Etienne brings to life the world of automobiles through his eloquent prose and an array of stunning, high-quality color photographs. His works are a tribute to the industry, capturing its evolution, technological advancements, and the sheer beauty of vehicles in a manner that is both informative and visually captivating.

A proud alumnus of the University of Malta, Etienne's academic background lays a solid foundation for his meticulous research and factual accuracy. His education has not only enriched his writing but has also fueled his career as a dedicated teacher. In the classroom, just as in his writing, Etienne strives to inspire, inform, and ignite a passion for learning.

As a teacher, Etienne harnesses his experience in writing to engage and educate, bringing the same level of dedication and excellence to his students as he does to his readers. His dual role as an educator and author makes him uniquely positioned to understand and convey complex concepts with clarity and ease, whether in the classroom or through the pages of his books.

Through his literary works, Etienne Psaila continues to leave an indelible mark on the world of automotive literature, captivating car enthusiasts and readers alike with his insightful perspectives and compelling narratives.

Visit www.etiennepsaila.com for more.

www.ingramcontent.com/pod-product-compliance
Ingram Content Group UK Ltd.
Pitfield, Milton Keynes, MK11 3LW, UK
UKHW020639161225
9592UKWH00003B/104